Christoph Strauss

**Precision spectroscopy with ultracold Rubidium molecules**

Christoph Strauss

# Precision spectroscopy with ultracold Rubidium molecules

## A detailed study of the triplet ground state and the first excited triplet state

Südwestdeutscher Verlag für Hochschulschriften

**Impressum/Imprint (nur für Deutschland/only for Germany)**
Bibliografische Information der Deutschen Nationalbibliothek: Die Deutsche Nationalbibliothek verzeichnet diese Publikation in der Deutschen Nationalbibliografie; detaillierte bibliografische Daten sind im Internet über http://dnb.d-nb.de abrufbar.
Alle in diesem Buch genannten Marken und Produktnamen unterliegen warenzeichen-, marken- oder patentrechtlichem Schutz bzw. sind Warenzeichen oder eingetragene Warenzeichen der jeweiligen Inhaber. Die Wiedergabe von Marken, Produktnamen, Gebrauchsnamen, Handelsnamen, Warenbezeichnungen u.s.w. in diesem Werk berechtigt auch ohne besondere Kennzeichnung nicht zu der Annahme, dass solche Namen im Sinne der Warenzeichen- und Markenschutzgesetzgebung als frei zu betrachten wären und daher von jedermann benutzt werden dürften.

Verlag: Südwestdeutscher Verlag für Hochschulschriften GmbH & Co. KG
Heinrich-Böcking-Str. 6-8, 66121 Saarbrücken, Deutschland
Telefon +49 681 37 20 271-1, Telefax +49 681 37 20 271-0
Email: info@svh-verlag.de

Approved by: Ulm, Uni, Diss., 2010

Herstellung in Deutschland:
Schaltungsdienst Lange o.H.G., Berlin
Books on Demand GmbH, Norderstedt
Reha GmbH, Saarbrücken
Amazon Distribution GmbH, Leipzig
**ISBN: 978-3-8381-1547-4**

**Imprint (only for USA, GB)**
Bibliographic information published by the Deutsche Nationalbibliothek: The Deutsche Nationalbibliothek lists this publication in the Deutsche Nationalbibliografie; detailed bibliographic data are available in the Internet at http://dnb.d-nb.de.
Any brand names and product names mentioned in this book are subject to trademark, brand or patent protection and are trademarks or registered trademarks of their respective holders. The use of brand names, product names, common names, trade names, product descriptions etc. even without a particular marking in this works is in no way to be construed to mean that such names may be regarded as unrestricted in respect of trademark and brand protection legislation and could thus be used by anyone.

Publisher: Südwestdeutscher Verlag für Hochschulschriften GmbH & Co. KG
Heinrich-Böcking-Str. 6-8, 66121 Saarbrücken, Germany
Phone +49 681 37 20 271-1, Fax +49 681 37 20 271-0
Email: info@svh-verlag.de

Printed in the U.S.A.
Printed in the U.K. by (see last page)
**ISBN: 978-3-8381-1547-4**

Copyright © 2012 by the author and Südwestdeutscher Verlag für Hochschulschriften GmbH & Co. KG and licensors
All rights reserved. Saarbrücken 2012

# Contents

| | | |
|---|---|---|
| | **List of Tables** | **3** |
| **1** | **Introduction** | **5** |
| | 1.1 Outline of the book | 7 |
| | 1.2 Publications | 8 |
| **2** | **Overview of the experimental setup** | **9** |
| **3** | **Production of Feshbach molecules** | **13** |
| | 3.1 Magneto-optical trap and optical molasses | 13 |
| | 3.2 Spin preparation | 16 |
| | 3.3 Magnetic transport | 17 |
| | 3.4 QUIC trap and evaporative cooling | 17 |
| | 3.5 Ultracold atoms in an optical lattice | 19 |
| | 3.6 Spin flip before Feshbach association | 22 |
| | 3.7 Ultracold Feshbach molecules in an optical lattice. | 24 |
| **4** | **Theory of diatomic molecules** | **27** |
| | 4.1 Some notation: From Operators to exact and good quantum numbers | 27 |
| | 4.2 Nonrelativistic Born-Oppenheimer and adiabatic approximation | 28 |
| | 4.3 Treatment with electronic and nuclear spins | 30 |
| | 4.4 Hund's cases without nuclear spin | 32 |
| | 4.5 Hund's coupling cases including nuclear spin | 35 |
| | 4.5.1 Case a) with basis functions | 35 |
| | 4.5.2 Case c) with basis functions | 38 |
| | 4.5.3 Case b) with basis functions | 38 |
| | 4.5.4 Case e) with basis functions | 39 |
| | 4.6 Symmetries | 39 |
| | 4.6.1 Exchange symmetries and Feshbach molecules | 41 |
| | 4.6.2 Molecular symmetries | 43 |
| | 4.7 Symmetries and Hund's cases in selected $^{87}Rb_2$ states | 45 |
| | 4.7.1 The electronic ground state connecting to the 5S+5S asymptote | 45 |
| | 4.7.2 The first electronic excited $(1)\,^3\Sigma_g^+$ state | 46 |
| | 4.8 Selection rules for bound-bound transitions | 48 |
| | 4.8.1 Derivation of the selection rules including nuclear spin | 50 |

## Contents

**5 One-photon spectroscopy of the $(1)\,^3\Sigma_g^+$ potential**    **55**
   5.1 Experimental setup . . . . . . . . . . . . . . . . . . . . . . . . . . 56
        5.1.1 Stability of lasers and uncertainties . . . . . . . . . . . . . . 59
   5.2 Experimental observations . . . . . . . . . . . . . . . . . . . . . . 60
        5.2.1 The splitting of the vibrational levels into $0_g^-$ and $1_g$ components . 61
        5.2.2 The spectra of the $0_g^-$ and $1_g$ states . . . . . . . . . . . . . . . 63
   5.3 Effective Hamiltonian and evaluation of molecular parameters . . . . 66
        5.3.1 Direct spin-spin and second-order spin-orbit interaction . . . . . . 66
        5.3.2 The hyperfine interaction . . . . . . . . . . . . . . . . . . . . . 68
        5.3.3 The Zeeman interaction . . . . . . . . . . . . . . . . . . . . . 69
        5.3.4 The rotational interaction . . . . . . . . . . . . . . . . . . . . 69
        5.3.5 The spin-rotational interaction . . . . . . . . . . . . . . . . . . 70
        5.3.6 Fit procedure and evaluation of molecular parameters . . . . . . . 70
   5.4 Discussion of results . . . . . . . . . . . . . . . . . . . . . . . . . 71
        5.4.1 $0_g^-$ spectrum and magnetic shifts . . . . . . . . . . . . . . . . 71
        5.4.2 $1_g$ spectrum and magnetic shifts . . . . . . . . . . . . . . . . 74

**6 Precision spectroscopy of the $a\,^3\Sigma_u^+$ potential**    **79**
   6.1 Experimental setup and dark state spectroscopy . . . . . . . . . . . . 81
   6.2 Quantum numbers and assignment . . . . . . . . . . . . . . . . . . 84
        6.2.1 Quantum numbers . . . . . . . . . . . . . . . . . . . . . . . 84
        6.2.2 Assignment of spectral lines . . . . . . . . . . . . . . . . . . 86
   6.3 Coupled-channel model and optimization of the $a\,^3\Sigma_u^+$ potential . . . . 89
   6.4 Progression of vibrational levels and their substructure . . . . . . . . . 94
        6.4.1 Vibrational ladder and rotational progression . . . . . . . . . . 94
        6.4.2 Hyperfine splitting and singlet-triplet mixing . . . . . . . . . . . 97
        6.4.3 Franck-Condon overlap $(1)\,^3\Sigma_g^+(v'=13) \to a\,^3\Sigma_u^+(v)$ . . . . . . . . 101
   6.5 Final analysis and reassignment of Feshbach resonances . . . . . . . . 102

**7 Summary and Outlook**    **105**

**8 Danksagung**    **109**

## List of Tables

| | | |
|---|---|---|
| 4.1 | Definitions of angular momentum operators and their quantum numbers. | 28 |
| 4.2 | Hund's coupling cases without nuclear spin. Good quantum numbers and nomenclature. | 34 |
| 4.3 | Atomic basis versus molecular basis: Clebsch-Gordon coefficients. | 43 |
| 4.4 | Symmetry of wave-functions with respect to the exchange of the nuclei. | 44 |
| 4.5 | Rotational ladder with symmetries for the electronic ground-state. | 46 |
| 4.6 | Rotational ladder with symmetries for the electronic excited $(1)\,^3\Sigma_g^+$ state. | 48 |
| 5.1 | Rotational structure with quantum numbers for the $0_g^-(v'=13)$ state. | 73 |
| 5.2 | Rotational structure with quantum numbers for the $1_g(v'=13)$ state. | 77 |
| 6.1 | Optimized potential for the $X\,^1\Sigma_g^+$ ground-state. | 95 |
| 6.2 | Optimized potential for the $a\,^3\Sigma_u^+$ ground-state. | 96 |
| 6.3 | Scattering lengths for the isotopic combinations $^{87}$Rb / $^{87}$Rb, $^{85}$Rb / $^{87}$Rb, and $^{85}$Rb / $^{85}$Rb. | 103 |

# 1 Introduction

There has been a quest for cold trapped atoms and molecules since the beginning of precision spectroscopy in the early stages of the 20th century. One of the most exciting questions concerned a quantum phase transition, the Bose-Einstein condensation (BEC) which was expected to occur for Bosons at temperatures below $1\,\mu$K. Another interesting field was high precision spectroscopy. In experiments with vapor cells, warm atoms or molecules move at high speeds which causes broadened and shifted spectral lines. Using atomic or molecular beams, one is able to enter the mK regime in the direction perpendicular to the beam but then the observation time is limited. The problem of broadened lines was solved using Doppler free saturation spectroscopy [Demtröder 2008] but the atoms are neither trapped nor cooled in these experiments. In 1975, T. Hänsch and A. Schawlow proposed a scheme where neutral atoms are directly cooled using the momentum transfer of laser light [Hänsch 1975][1]. The breakthrough in laser cooling came in the 1980s when alkali-atoms were cooled to the Doppler-limit [Chu 1985]. It was a great surprise that the atoms could be cooled below the Doppler-limit [Lett 1988, Dalibard 1989] (for a review article see [Adams 1997]). Unfortunately, compared to typical temperatures where the transition to a BEC takes place, the atoms were still too hot. A promising candidate for further cooling was evaporative cooling [Ketterle 1999a], where the hottest atoms are thrown away in a controlled way using radio-frequency. Almost twenty years later, BEC with $^{87}$Rb [Anderson 1995], $^{23}$Na [Davis 1995] and $^{7}$Li [Bradley 1995] atoms was achieved. Shortly afterwards, several key experiments were carried out showing the quantum nature of the condensate, for example, the interference between two condensates showing the coherence in the condensate [Andrews 1997]. Henceforward, a rush in the field began.

But what about cold molecules? The laser-cooling schemes, developed for atoms, are difficult to implement because molecules get lost quickly as they have more degrees of freedom than atoms. First experiments with cold molecules started from a magneto-optical trap using photoassociation spectroscopy [Lett 1993, Takekoshi 1998, Weiner 1999]. For recent review articles, see [Hutson 2006, Jones 2006, Carr 2009]. Other techniques use supersonic beams [Bahns 1996] or Helium nanodroplet isolation [Toennies 1998]. Although not obvious, Helium nanodroplet isolation is also based on supersonic beams and the translational temperatures in these experiments are high. Thus, it is difficult to trap or even further cool the pre-cooled molecules. In 2003, a different association scheme entered the field: bosonic [Herbig 2003] and fermionic molecules [Regal 2003] were created using a Feshbach resonance. Such a scattering resonance appears if a state consisting of

---

[1] At the same time D. Wineland and H. Dehmelt suggested a cooling scheme for ions [Wineland 1975]. In the following, I will focus on cold atoms, because ions are not of interest in our experiments.

# 1. Introduction

two atoms becomes energetically degenerate with a molecular state. In the regime close to degeneracy, a resonance in the interaction strength (that is the scattering length) as a function of the magnetic bias field is observed. Creating molecules using a Feshbach resonance became a standard procedure, and is applicable to a lot of homo- and heteronuclear species (For review articles, see [Köhler 2006, Chin 2010].). In all these experiments, the Feshbach association starts from an atomic gas with high phase-space density and results in the most weakly bound molecules. This means that the molecules have the highest vibrational quantum number and binding energies are on the order of a few MHz×$h$, where $h$ is Planck's constant. Early experiments revealed that weakly bound bosonic dimers are not stable due to inelastic collisions [Xu 2003]. In contrast, molecules consisting of two fermionic atoms are comparatively more long-lived [Petrov 2004], and, as mentioned above, the interaction between the atoms can be tuned using the magnetic field. This "knob" has made it possible to reach a molecular Bose-Einstein condensate consisting of molecules which are composed of two fermionic atoms in the limit of vanishing binding energy [Greiner 2003, Jochim 2003, Zwierlein 2003, Bourdel 2004]. As already mentioned, there was a problem producing long-lived samples of bosonic molecules. This problem was solved using three mutually orthogonal laser beams which form an optical lattice. In the case of Feshbach molecules, an optical lattice allows Feshbach association with almost unit efficiency in well defined quantum states [Thalhammer 2006]. In this experiment Thalhammer *et al.* also demonstrated that the optical lattice leads to long-lived Feshbach molecules. Shortly after, it was also shown that the association leads to a molecular Mott insulator state where an occupation of exactly one molecule per lattice site is retrieved [Volz 2006]. In an interesting experiment, Winkler *et al.* showed, that a coherent transfer of these weakly bound molecules to a more deeply bound state is possible with almost unit efficiency [Winkler 2007a]. This scheme made use of a coherent transfer scheme [Fewell 1997, Bergmann 1998], and it seemed possible that it could be extended to arbitrary states, provided one had a molecular lambda system with corresponding lasers. Thus, if the molecular structure is known, one could use a Raman transition to transfer the molecules to the ro-vibrational ground state. These ground state molecules are the starting point for theory proposals. For example, one can lower the lattice potential after the production of ground-state molecules to achieve a BEC of molecules in the ro-vibrational ground state [Jaksch 2002, Damski 2003].

Another set of interesting experiments concerns heteronuclear molecules [Ospelkaus 2006] which possess a permanent dipole moment. In this case, the dipole-dipole interaction is long-range and spatially anisotropic, which is in contrast to the short-range van der Waals interactions for homonuclear molecules. These molecules were produced recently [Ni 2008, Ni 2009, Aldegunde 2008], and are proposed to serve as an interesting tool for quantum simulation of condensed matter spin systems [Pupillo 2008], or for quantum information processing schemes [DeMille 2002, Andrè 2006, Yelin 2006].

Recently, optical schemes have been developed to selectively produce cold and dense samples of deeply bound molecules in well defined quantum states [Danzl 2008], [Lang 2008, Ni 2008, Ospelkaus 2010] using a Raman transition. Additionally, other non-coherent schemes were developed. These schemes involve femtosecond lasers for optical pumping [Viteau 2008], and photoassociation of ultracold LiCs molecules in their ro-vibrational ground state [Deiglmayr 2008]. All of these (non-) coherent schemes have opened up new possibilities for cold collision experiments [Ospelkaus 2010, Ni 2010], ultracold chemistry [Staanum 2006, Krems 2005, Krems 2008], and for testing fundamental laws via precision spectroscopy [Doyle 2004, Flambaum 2007, Zelevinsky 2008, DeMille 2008, Chin 2009]. For such future experiments it is mandatory that the location and properties of the available molecular quantum states are well known and understood. The triplet ground state, for example, has been largely unexplored, as the molecules are normally found in their singlet states. From these states, a dipole transition to triplet states is optically forbidden due to the common selection rule $\Delta S = 0$ where we denote the total electronic spin $S$.

There has been a great progress in the development of different cooling schemes in the meantime. Here, I want to mention experiments with molecules embedded in Helium nanodroplet beams (For a review see [Stienkemeier 2006].), and experiments with molecular beams (See review articles by [Van de Meerakker 2008, Schnell 2009]), but ultracold, trapped molecules remain challenging [Dulieu 2009]. Recently, it even became possible to directly laser-cool molecules optically for the first time [Rosa 2004, Shuman 2010]. This again shows that the field of ultracold molecules is a hot topic at the moment.

## 1.1 Outline of the book

The book is organized as follows: In chapter 2, I give an overview of the hardware of the experimental setup, including the vacuum chambers, the laser system and the computer control of the experiment.
Chapter 3 reviews the different steps which we use to produce ultracold $^{87}$Rb$_2$ in an optical lattice. Here, I will explain in detail how we create the molecules from $^{87}$Rb atoms in a Magneto optical trap.

Chapter 4 gives a brief introduction to the theory of homonuclear diatomic molecules. I will start with the Born-Oppenheimer approximation. The strength of direct spin-spin, spin-orbit or hyperfine interactions is classified in Hund's coupling schemes which I introduce consecutively. At the end of the chapter, I will summarize the symmetries which are important for the discussion of selection rules.

Chapters 5 and 6 are dedicated to the experimental results: I first present one-photon spectroscopy, where we use a laser to couple the Feshbach molecules to the first excited $(1)\,^3\Sigma_g^+$ state. For a better understanding of the molecular structure, we also varied the magnetic field to get information on the Zeeman shift of the excited state. This has been done using "adiabatic transfers over avoided crossings" as introduced by [Lang 2008a]. At the end, we compare our measurements with numerical simulations done by Marius Lysebo and Leif Veseth in Oslo involving effective Hamiltonians to obtain an understanding of the molecular structure.

Our Raman spectroscopy is discussed in chapter 6. Using dark-state spectroscopy, we resolve vibrational, rotational, hyperfine and Zeeman structure. We can directly observe singlet-triplet mixing at binding energies as high as a few hundred GHz. Furthermore, we studied the dependence of the hyperfine, spin-spin, and rotational interactions on the vibrational quantum number. With the help of Eberhard Tiemann, who analysed our results theoretically using a coupled channel model we are able to fully understand the observed structure.

The book ends with a short summary and an outlook towards further exciting experiments with deeply bound triplet molecules.

## 1.2 Publications

This book is based on my PhD thesis [Strauss2011]. The results presented in this book were also published in the following articles:

- Hyperfine, rotational, and vibrational structure of the $a^3\Sigma_u^+$ state of $^{87}Rb_2$.
  C. Strauss, T. Takekoshi, F. Lang, K. Winkler, R. Grimm, E. Tiemann, and J. Hecker Denschlag,
  Phys. Rev. A **82**, 052514 (2010).

- Hyperfine, rotational and Zeeman structure of the lowest vibrational levels of the $^{87}Rb_2$ (1) $^3\Sigma_g^+$ state.
  T. Takekoshi, C. Strauss, F. Lang, J. Hecker Denschlag, M. Lysebo, and L. Veseth,
  Phys. Rev. A **83**, 062504 (2011).

Closely related work on the production and dynamics of ultracold $^{87}Rb_2$ molecules can be found in:

- Cruising through molecular bound-state manifolds with radiofrequency.
  F. Lang, P.v.d. Straten, B. Brandstätter, G. Thalhammer, K. Winkler, P.S. Julienne, R. Grimm, and J. Hecker Denschlag,
  Nature Phys. **4**, 223 (2008).

- Ultracold Triplet Molecules in the Rovibrational Ground State.
  F. Lang, K. Winkler, C. Strauss, R. Grimm, and J. Hecker Denschlag,
  Phys. Rev Lett. **101**, 133005 (2008).

- Dark state experiments with ultracold, deeply-bound triplet molecules.
  F. Lang, C. Strauss, K. Winkler, T. Takekoshi, R. Grimm, and J. Hecker Denschlag,
  Faraday Discussions **142**, 271-282 (2009).

## 2 Overview of the experimental setup

This experiment was started in 2001 in Innsbruck at the "Institut für Experimentalphysik" (Austria) and moved to the "Institut für Quantenmaterie" in Ulm (Germany) in 2009. The main feature is a magnetic transport line [Greiner 2000, Greiner 2001] which connects the magneto optical trap (MOT) chamber with the glass cell (Fig. 2.1). The idea behind the magnetic transport is to separate the chamber where we create the MOT from another chamber (glass cell) with excellent optical access and ultra high vacuum. Furthermore, the glass cell has the advantage of negligible magnetic permeability. Thus eddy currents are not a problem when we switch on currents up to 100 A, which corresponds to magnetic fields up to 1000 G.

**Experimental table.** We need ultra high vacuum in the chambers because our atoms and molecules are lost from the trap when they collide with gas from the background vapor. The chambers are located on an optical table and are surrounded by optical elements to guide the laser beams for trapping and cooling of the atoms or molecules. The experimental cycle starts by trapping atoms in a magneto optical trap (MOT) from the background vapor, where we have a vacuum of better than $10^{-8}$ mbar. This chamber is connected to an ion getter pump and a $^{87}$Rb reservoir (compare Fig. 2.1). At room temperature we have a $^{87}$Rb pressure of $\approx 2 \times 10^{-7}$ mbar which is sufficient for daily operation. The $^{87}$Rb reservoir can be separated from the MOT chamber with the aid of a valve for exchanging the $^{87}$Rb source.

We use a sequence of magnetic quadrupole traps to transfer the cold atoms to the glass cell. During the transfer the atoms are always kept at the minimum of the field which we create with pairs of coils mounted below and above a tube. This tube has an inner diameter of 6.2 mm and is used as differential pumping stage. It ensures a vacuum of better than $10^{-11}$ mbar in the glass cell [Theis 2005]. The glass cell itself is pumped with a second ion getter pump and a Titanium sublimation pump. A detailed description of the vacuum system and the magnetic transport line can be found in the diploma thesis of Klaus Winkler [Winkler 2002] and the doctoral thesis of Matthias Theis [Theis 2005].

Furthermore, it is necessary to control all field currents at a high level of accuracy in order to transfer and manipulate the atoms magnetically in an efficient way. We have the possibility to use the power supplies (Delta-Electronica) either in a constant current mode (CC) or a constant voltage mode (CV). In the CC mode the current is regulated by an internal control loop which has the disadvantage of increased noise in current by a factor of ten compared to the CV mode. If we need very precise currents and fast switching we use the power supplies in CV mode with an external control loop. The external digital PID-regulator then reaches a precision of $4 \times 10^{-5}$ [Thalhammer 2007].

## 2. Overview of the experimental setup

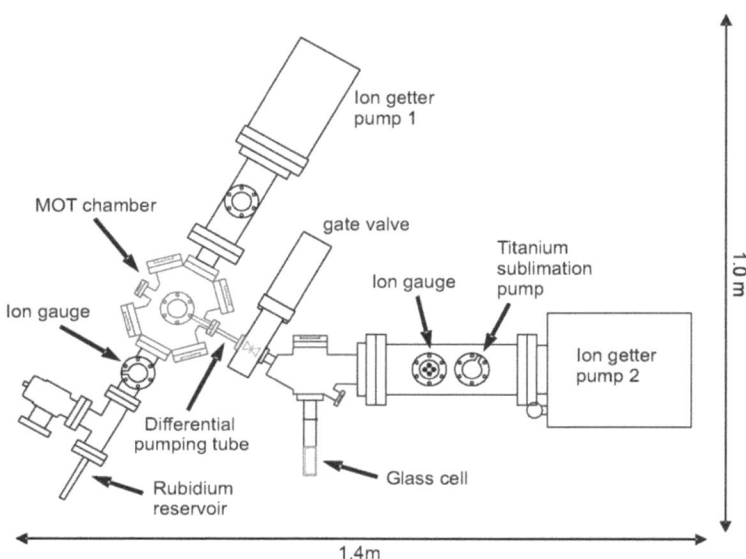

Figure 2.1: Vacuum system on the optical table. The Rubidium reservoir, which is at room temperature, is next to the MOT chamber. We connect this chamber on the right hand side via a differential pumping stage to the glass cell. The MOT chamber is pumped with an ion getter pump to get a vacuum of better than $10^{-8}$ mbar. The glass cell is normally also pumped with an ion getter pump and a Titanium sublimation pump. The pressure in the glass cell is better than $10^{-11}$ mbar. Adapted from [Winkler 2007].

## 2. Overview of the experimental setup

**Laser table.** A further crucial element in the setup is the laser system. The lasers are located on separate optical tables. We have to stabilize the frequency of the lasers to manipulate the $^{87}$Rb atoms. Therefore, we first lock the "Master laser" to an atomic $^{87}$Rb line at 780.25 nm using modulation transfer spectroscopy [Bjorklund 1983], [Thalhammer 2007]. See [Steck 2003] for data on atomic $^{87}$Rb. Afterwards most of the other lasers are locked to this Master laser. Typical locking schemes which we use are the beat-lock [Schuenemann 1999] (seed laser for MOT-light, cavity reference laser, and Kick-laser for purification scheme in the optical lattice), and the Pound-Drever-Hall (PDH) lock [Drever 1983, Black 2001], which is used to lock a reference cavity. We create the MOT light with a home built diode laser at 780 nm. Afterwards we amplify the light with a home built tapered amplifier using an injection lock. Some light from the Master laser is also used for absorption imaging [Ketterle 1999]. Unfortunately, the MOT and absorption light also excites some of the atoms to the $5P_{3/2}$, $f' = 2$ state which can decay to the $5S_{1/2}$, $f = 1$ state (See also Fig. 3.8 at page 24.). These atoms are lost because they cannot be excited anymore with the MOT or absorption light; they are in a dark state. In order to complete the cycling transition for $^{87}$Rb atoms, we use a repumper laser at 780.23 nm driving the $|5^2 S_{1/2}, f = 1\rangle \rightarrow |5^2 P_{3/2}, f' = 2\rangle$ transition. We lock this laser separately using frequency-modulation spectroscopy. It also serves as a reference to estimate the accuracy and long time stability of our wave meter (WS7 Highfinesse). We connect all beams via optical fibers to the experimental table.

We use a Verdi V18 (Coherent) with an output power of 18.5 W at 532 nm to pump two Ti:sapphire lasers. The first of these lasers has an output power of ≈300 mW at 830.44 nm and is used for the optical lattice. We use a second Ti:sapphire laser for spectroscopy and for the transfer of molecules to more deeply bound levels (Chapter 6). If the Lyot filter and the two Etalons are mounted, it has an output power ranging from 100 mW at 1060 nm to 1 W at 980 nm. The laser for the lattice is always free-running, whereas we can lock the spectroscopy laser if necessary using a PDH lock to a cavity. Another diode laser from Toptica (DL100 Pro) which has a maximum power of 50 mW at 1050 nm (32 mW at 985 nm to 24 mW at 1066 nm) is also used for spectroscopy or for the transfer of molecules to more deeply bound levels.

**Software to control the experiment.** The user interface is written in Labview. The calculations and the evaluation of the pictures are done with Matlab to speed up the experimental cycle. A Python server controls all the steps and calls an external processor (Adwin Gold) and all other devices, for example radio-frequency synthesizers. The computer hardware and the computer software is well explained in the doctoral thesis of Gregor Thalhammer [Thalhammer 2007].

# 3 Production of Feshbach molecules

The starting point for all experiments is a pure sample of Feshbach molecules, trapped in an optical lattice. The steps to obtain the pure sample are

- the creation of a magneto-optical trap with subsequent optical molasses
- loading the atoms into the first magnetic trap followed by magnetic transport to the glass cell
- transfer of the atoms from a magnetic trap to a quadrupole-Ioffe-configuration (QUIC) trap
- evaporative cooling of the pre-cooled sample of $^{87}$Rb atoms down to the critical temperature
- loading the cloud of ultracold atoms into the optical lattice
- Feshbach association and purification in the lattice.

I will review the basic steps which lead to the pure sample of Feshbach molecules in the following. This is described in more detail in the theses of Gregor Thalhammer and Matthias Theis [Thalhammer 2007, Theis 2005].

## 3.1 Magneto-optical trap and optical molasses

The experimental cycle starts with the magneto-optical trap (MOT), where we load approximately $3 \times 10^9$ $^{87}$Rb atoms from the background vapor. The loading takes about 8 s and we reach a final temperature of about 150 $\mu$K. These numbers stem from fits to absorption images as explained in [Theis 2005]. The light for trapping originates from a home-built tapered amplifier with an output power of about 50 mW in each beam. This laser is seeded by a home-built diode laser. It is shifted 20 MHz from the atomic D$_2$ $|5^2S_{1/2}, f = 2\rangle \rightarrow |5^2P_{3/2}, f' = 3\rangle$ resonance of $^{87}$Rb as explained in chapter 2. Here we use the total atomic angular momentum $f = l + s + i$ including the total orbital angular momentum $l$, all electronic ($s$) and nuclear ($i$) spins[1]. We use a second home-built diode laser to complete the cycling transition. In the literature, this laser is referred to as a repumper [Metcalf/van der Straten 1999]. This repumper laser is locked to the atomic $|5^2S_{1/2}, f = 1\rangle \rightarrow |5^2P_{3/2}, f' = 2\rangle$ transition. The basic principle of the magneto-optical

---

[1] In the following, small letters $f$, $l$, ... refer to atomic spins whereas capital letters $F$, $L$, ... refer to spins in the $^{87}$Rb$_2$ molecule. A level scheme for the atoms can be found in Fig. 3.8 at page 24.

## 3. Production of Feshbach molecules

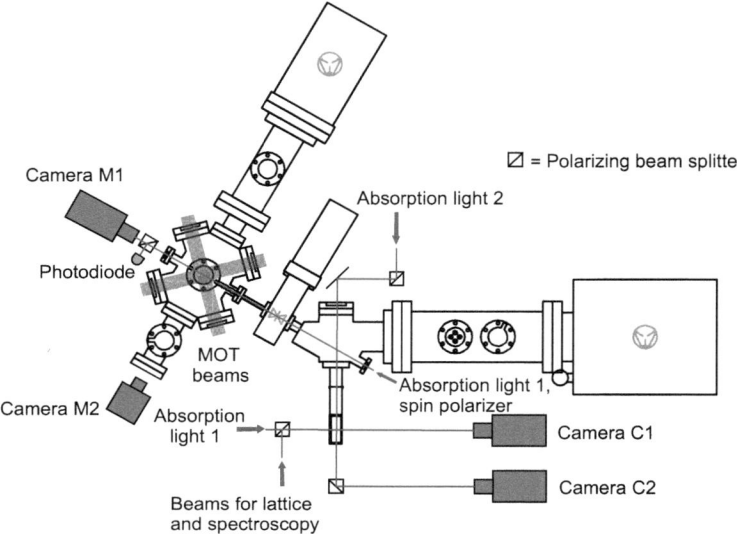

Figure 3.1: Experimental setup including laser beams. The figure shows the MOT beams and the imaging setup. We have two cameras (M1 and M2) to get fluorescence or absorption images in the MOT. In the glass cell we perform absorption imaging with the cameras C1 and C2. A camera to get pictures in the glass cell in the third direction is not shown. The light which we use for absorption imaging is called "Absorption light 1 and 2". Adapted from [Theis 2005].

trap is Doppler cooling. This process has a lower bound of [Adams 1997]

$$k_B T_D = \frac{\hbar \cdot \Gamma}{2} \quad (3.1)$$

caused by the lifetime of the excited state and Heisenberg's uncertainty principle. This temperature results from the competition between a viscous cooling force and a random heating force. A careful analysis shows that the heating is due to spontaneous emission. In Eq. (3.1), $T_D$ is the Doppler temperature, which is on the order of $100\,\mu K$ for $^{87}$Rb. $1/\Gamma$ is the lifetime of the excited state. The principle of the MOT can be understood as follows. The photon energy is set in such a way that only atoms which move at a special velocity towards the beam absorb a photon (red detuned MOT-laser) and receive a recoil. The effect of cooling appears because the incident laser light is directed whereas the fluorescence is not. In contrast, the fluorescence of the excited atom has no preferred direction. Taking the average of several absorption-emission cycles gives a net

3. Production of Feshbach molecules

Figure 3.2: Magnetic transport line. The Figure shows the coils which we use in our setup. The yellow coil on the left hand side creates the magnetic quadrupole field for the MOT, the green coil on the left hand side is used to push the atoms towards the first magnetic trap of the transport line (blue coils). The transport line consists in total of 2 × 13 coils (blue, red and yellow). The glass cell together with the three coils which are also used for the QUIC trap (yellow) are at the end of the transport line on the right hand side. Adapted from [Winkler 2007].

momentum transfer and causes cooling [Hänsch 1975]. The atoms are also trapped and always in resonance with the cooling light due to a magnetic offset field. This field shifts the different $m_f$ levels, where we use the projection of $f$ in direction of the magnetic field. In our experiment we use a combination of magnetic fields and circular polarized light, which leads to a force directed to the center of the trap. However, the theory is more complicated as one has to take the hyperfine-structure of the $^{87}$Rb atoms into account [Metcalf/van der Straten 1999, Foot 2005].

Afterwards we start the molasses phase where we switch the repumper laser and the magnetic fields off. At the same time we increase the detuning of the MOT light. After 15 ms our atoms have a temperature of about 50 $\mu$K. This sub-doppler cooling can be understood when we include the polarizations of the laser beams, which do not play a major role in Doppler cooling. Unfortunately, the theory in three dimensions is rather complicated and no analytical solutions exist. Therefore, we consider the one-dimensional case to get a feeling for the physics [Dalibard 1989]. In this case, the lower-temperature bound of the process in the low-intensity regime is given as [Metcalf/van der Straten 1999]

$$k_B T_{SD} \propto \frac{\delta \cdot I}{1 + \left(\frac{2\delta}{\Gamma}\right)^2}, \qquad (3.2)$$

where we use the intensity $I$ in each laser beam, the detuning $\delta$ from the atomic resonance, and the lifetime $1/\Gamma$ of the excited atomic state. In principle Eq. (3.2) shows that we can lower the temperature by increasing the detuning and lowering the power

3. Production of Feshbach molecules

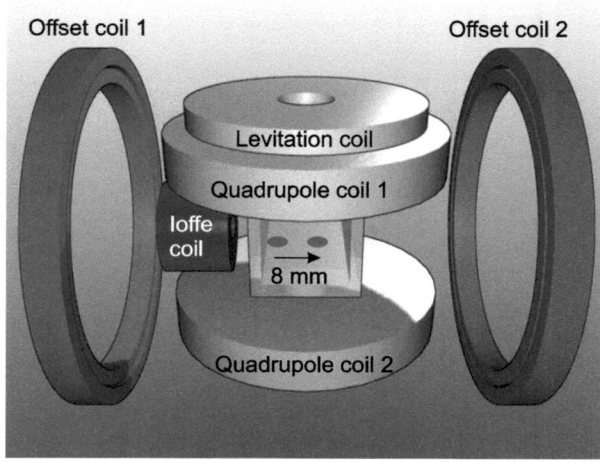

Figure 3.3: Configuration of the coils near the glass cell. The levitation coil creates an inhomogeneous magnetic field and is used to compensate for the gravitational force. We use the two quadrupole coils 1 and 2 for the Feshbach field as well as for the QUIC trap. The Ioffe coil is added to form the QUIC trap. Unfortunately, the center of the QUIC trap is shifted 8 mm with respect to the center of the quadrupole coils. Therefore, we have to move the cloud of atoms (red) with the offset coils into the center of the Feshbach field. Adapted from [Winkler 2007].

of the cooling laser. For details I refer the reader to the excellent article by Dalibard [Dalibard 1989] and the review written by [Adams 1997]. The absolute lower-temperature bound in this scheme is the recoil limit with

$$k_B T_r = \frac{\hbar^2 k^2}{M}. \tag{3.3}$$

Here $k$ is the wave vector of the light field, $2\pi\hbar$ is Planck's constant, and $M$ is the mass of the $^{87}$Rb atom. This limit arises as the spontaneous emission of light at some point contributes to the heating process. After the molasses phase we end up with $3 \times 10^9$ $^{87}$Rb atoms at temperatures as low as $50\,\mu$K.

## 3.2 Spin preparation

For our experiments we want to have atoms in the $5^2 S_{1/2}$ $|f = 1, m_f = -1\rangle$ state (See Fig. 3.8 at page 24 for the Zeeman structure.). In the MOT coils the $|f = 1, m_f = -1\rangle$,

$|f = 2, m_f = 1\rangle$, and $|f = 2, m_f = 2\rangle$ states are magnetically trapped. Thus, we have to pump as many atoms to the target state as possible. This transfer already starts when we switch off the repumper laser as it leads to a depopulation of the state $|5^2S_{1/2}, f = 2\rangle$. Additionally, we switch on a magnetic-field gradient of 14 G/cm which we create with the push-coil (green coil on the left in Fig. 3.2). At the same time we switch on $\sigma^-$ polarized light for 250 µs. This light is resonant with the transition $|5^2S_{1/2}, f = 1\rangle \rightarrow |5^2P_{3/2}, f' = 1\rangle$ and pumps the atoms in the desired $m_f = -1$ level. Unfortunately, spontaneous decay from the excited level also populates the $f = 2$ state. Further $\sigma^-$ polarized light which is resonant with $|5^2S_{1/2}, f = 2\rangle \rightarrow |5^2P_{3/2}, f' = 2\rangle$ completes the optical pumping process.

## 3.3 Magnetic transport

At the end of the molasses phase we again create a magnetic trap with the MOT coils which serves as starting point for the subsequent magnetic transport. During the transport we lose about 50% of our atoms and the temperature increases to 200 µK. For the magnetic transport we use in total 2 × 13 coils. They are shown in Fig. 3.2. The design is such that the local minimum of the magnetic field can be moved continuously over the whole transport line [Greiner 2000, Greiner 2001]. We have a magnetic-field gradient of 130 mG/cm to trap $^{87}$Rb atoms in the $| f = 1, m_f = -1 \rangle$ state. For the current ramps up to 100 A we use four power supplies from Delta electronica combined with home-built control and switch boxes. The control box only serves as a security device. The four switch boxes transfer the current from the associated power-supply to the different coils. This technique reduces the number of power supplies drastically. For the magnetic transport we use the power supplies in constant-current mode combined with a compensation of the frequency response as described by Winkler [Winkler 2007]. As already mentioned, the constant-current mode has the disadvantage of an increased current-noise by a factor of 10 which is not a problem at this point. The atoms are heated slightly during the 1.5 s which are needed for the transport. At the end of the transport sequence we have about $5 \times 10^8$ atoms at 200 µK in a stiff magnetic quadrupole trap (Yellow coils above and below the glass cell in Fig. 3.2. See also Fig. 3.3).

## 3.4 QUIC trap and evaporative cooling

The magnetic transport ends in the glass cell where we create a magnetic quadrupole trap using a current of ≈40 A. This field is not suitable for evaporative cooling because the magnetic field is zero in the center of the trap. At zero magnetic field the different $m_f$ levels are degenerate which causes Majorana flips and induces losses [Ketterle 1999a]. Therefore, we create a QUIC trap by ramping up a current through the Ioffe coil so that the total current through the quadrupole and Ioffe coils stays constant [Esslinger 1998]. The geometry of the QUIC coils can be seen in Fig. 3.2 where the Ioffe coil to the right

## 3. Production of Feshbach molecules

Figure 3.4: Production of a BEC. The peak height and the color represent the atomic density. The cloud on the left hand side is purely thermal and has a Gaussian shape. In the central picture one can see a thermal fraction of atoms with Gaussian momentum distribution and a small condensed cloud of atoms which has a Thomas-Fermi momentum profile. The picture on the right corresponds to a pure BEC of roughly $2 \times 10^6$ atoms at a temperature below 500 nK. Adapted from [Winkler 2007].

hand side is marked yellow (compare also Fig. 3.3). In the QUIC trap we have trapping frequencies of 16 Hz in the radial and 150 Hz in the axial direction. The formation of a cigar-shaped trap increases the density, and elastic scattering rates are enhanced. Evaporative cooling is more efficient [Foot 2005].

The Ioffe field causes an offset which prevents uncontrolled flips into different hyperfine states. During the creation of the QUIC trap, we only lose the hottest particles which cools the atom cloud. At the onset of the QUIC trap we have $4 \times 10^8$ atoms at temperatures below $500\,\mu$K. Then we begin an evaporation ramp from 30 MHz to 1.4 MHz. At the end of this ramp we obtain a pure BEC of $4 \times 10^6$ atoms in the $|f=1, m_f = -1\rangle$ state at temperatures of about 500 nK. The phase transition from a thermal cloud to a pure BEC is shown in Fig. 3.4. After the evaporation has finished we finally use the offset coils 1 and 2 to move the atoms back to the center of the magnetic trap. (For the coil geometry see Fig. 3.3.) This is necessary as we perform the Feshbach association in the magnetic field created by quadrupole coils 1 and 2. It turns out that a trapping frequency of 150 Hz at the end of the evaporation causes parametric heating [Thalhammer 2007]. This is due to noise at the third harmonic of the line frequency which is at 50 Hz. We therefore lower the magnetic field during the last seconds of evaporation. This leads to a reduced trapping frequency of 130 Hz in axial direction. The creation of the BEC is a crucial step when checking for proper operation of the experiment. The measured particle numbers and temperatures serve as a reference. For our measurements we use a thermal cloud close to condensation as more particles result in a better signal to noise ratio.

**Remark.** We need a spin flip to the state $|f=1, m_f = 1\rangle$ to create Feshbach molecules. For this spin flip we have additional coils (Not shown in Fig. 3.3.) which are mounted on a cage surrounding the glass cell. All currents through the quadrupole and Ioffe coils are designed such that this extra field is needed.

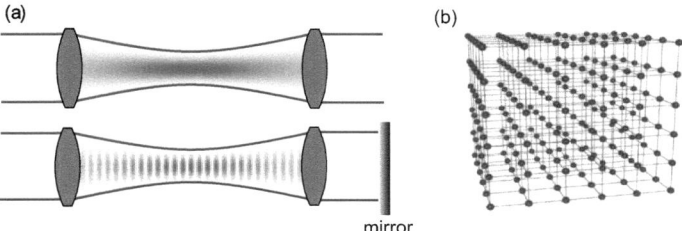

Figure 3.5: Lattice beam alignment. a) Rb atoms (red) in a focused beam. In the lower picture we show Rb atoms in a retro-reflected beam. Here the light creates a standing wave which corresponds to a periodic potential for the atoms. With three mutually orthogonal beams we create a cubic lattice as shown schematically in part b). Adapted from [Winkler 2007].

## 3.5 Ultracold atoms in an optical lattice

As a starting point for our experiments with ultracold molecules we choose an atomic cloud with a considerable fraction of thermal atoms and stop the evaporation before a pure condensate forms, which corresponds to radio frequencies of about 1.6 MHz. Stopping the evaporation at a slightly higher temperature ensures a larger number of atoms and Feshbach molecules. We then shut off all magnetic fields and ramp up a lattice which we create from a Ti:sapphire laser beam. Normally, we use 100 mW of light in each direction at a vacuum wavelength of 830.44 nm. The potential created by a focused retro-reflected beam is shown in Fig. 3.5. In this case the beam not only works as a dipole trap but also generates a standing wave, which induces a periodic potential for the atoms.

If we want to understand how the periodic structure for the atoms (or molecules) emerges we consider the simplified case of a lattice in one dimension, which I choose to be the $z$-direction. We take a laser beam which has a Gaussian intensity profile

$$I(r,z) = \frac{2P}{\pi \omega^2(z)} e^{-2\frac{r^2}{\omega^2(z)}}$$
$$\omega(z) = \omega_0 \sqrt{1 + (z/z_R)^2}. \tag{3.4}$$

Here $P$ denotes the total power in the beam, $\omega(z)$ is the $1/e^2$ intensity radius. $\omega_0$ is the radius at the focus $z = 0$ and $z_R = \frac{\pi \omega_0^2}{\lambda}$ is the Rayleigh length. The Rayleigh length is the distance at which the beam doubles its area compared to $z = 0$. Inserting Eq. (3.4)

## 3. Production of Feshbach molecules

into the formula for a potential created by the dipole force $V = -1/2 \cdot \langle \mathbf{d} \cdot \mathbf{E} \rangle$ leads to

$$V(r,z) = V_L e^{-2\frac{r^2}{\omega^2(z)}} \cos^2(k_L z)$$
$$\approx -V_L \left(1 - \frac{2r^2}{\omega_0^2}\right) \cos^2(k_L z) \quad (3.5)$$

after some calculation. Here, we have used the lattice depth $V_L$ and the wave vector $k_L = 2\pi/\lambda$. In our experiment we have $P = 100\,\mathrm{mW}$ in each direction, $\omega_0 = 130\,\mu\mathrm{m}$ which corresponds to a Rayleigh length of $500\,\mu\mathrm{m}$. Thus $\omega(z) \approx \omega_0$ and the approximation in Eq. (3.5) is justified. One advantage of reflected beams is that the lattice potential is four times deeper compared to that of a non-reflected beam. If we create the lattice with a laser which is detuned far[2] from the atomic resonance the following relation holds for the lattice depth [Grimm 2000]:

$$\frac{V_L}{E_r} = \frac{2m}{\hbar^2 k_L^2} \cdot \frac{3\pi c^2}{2\omega_0^3} \cdot \frac{\Gamma}{\Delta} \cdot I(r,z). \quad (3.6)$$

Here it is convenient to give $V_L$ in units of recoil energies $E_r = (\hbar k_L)^2/2m$ with the mass $m$ of one atom. The linewidth $\Gamma$ of the excited state is related to the dipole matrix element. $\Delta$ is the laser detuning from the center of the $D_1$ and $D_2$ line.

In the harmonic approximation of $I(r,z)$ we obtain trapping frequencies of

$$\omega_z = \sqrt{\frac{2V_L k_L^2}{m}} \quad (3.7)$$

in the axial and

$$\omega_r = \sqrt{\frac{4V_L}{m\omega_0^2}} \quad (3.8)$$

in the radial direction, with $m$ the mass of a $^{87}\mathrm{Rb}$ atom. As we have $\omega_z = \sqrt{k_L^2 \omega_0^2/2} \cdot \omega_r \gg \omega_r$ we can consider the lattice as a chain of micro traps with a trap frequency of $\omega_z$.

In our experiment we use a Ti:sapphire laser at $\lambda = 830.44\,\mathrm{nm}$ for the lattice. The linewidth of the excited state is $\Gamma/2\pi = 5.9\,\mathrm{MHz}$[3]. The recoil energy is $E_r = 2 \times 10^{-36}\,\mathrm{J}$. We ramp up the lattice adiabatically to the maximum available power corresponding to a depth of $37\,E_r$. On a temperature scale this corresponds to a trap depth of $V_L/k_B = 6\,\mu\mathrm{K}$. The beams are retro-reflected with mirrors of reflectivity 0.99. We use the transmitted light for an intensity lock using acousto optical modulators as described in the thesis of Thalhammer [Thalhammer 2007]. In the three dimensional lattice the interference between beams in different directions can lead to decoherence in the atomic sample. We choose the polarizations to be mutually orthogonal to avoid these interference effects. Furthermore, the frequencies of the different beams are detuned by about

---
[2]compared with the excited-state hyperfine splitting.
[3]Here we take the mean of the $D_1$ and $D_2$ lines as an approximation

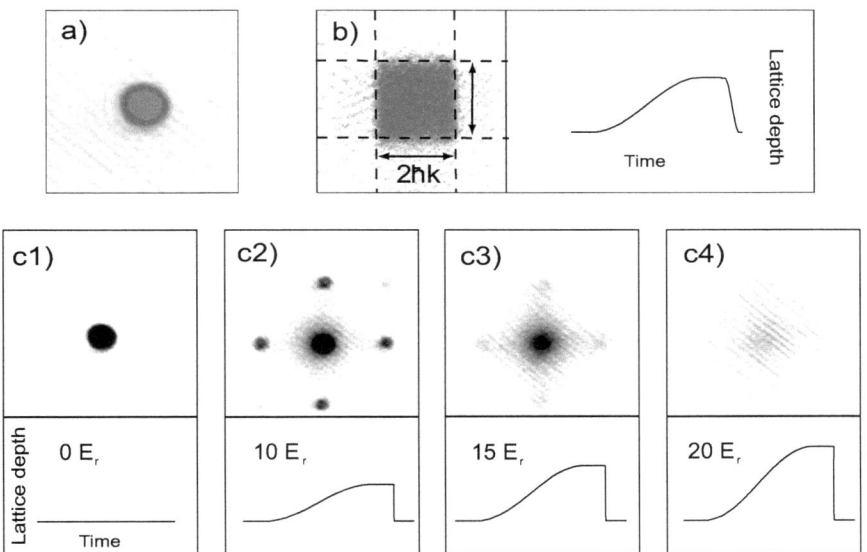

Figure 3.6: $^{87}$Rb atoms in an optical lattice. a) An absorption image of a pure BEC of $^{87}$Rb atoms after a time of flight of about 15 ms. b) Time of flight image for atoms which are released from an optical lattice. The lattice has a depth of 35 $E_r$ and is switched off adiabatically during 3 ms. The momentum width $2\hbar k$ of the square corresponds to the first Brioullin zone and is 175 $\mu$m wide. The sequence c1) to c4) shows the disappearance of the interference pattern. As explained in the text, c2) corresponds to a superfluid state, whereas c4) shows the Mott-insulator state. Picture c3) corresponds to an intermediate case. Adapted from [Thalhammer 2007].

## 3. Production of Feshbach molecules

100 MHz using an AOM. Any remaining interference terms in the potential thus average out [Greiner 2003a]. Fig. 3.6 a) shows an absorption image of an expanded ultracold cloud of $^{87}$Rb atoms close to degeneracy. For Fig. 3.6 b) we use the same experimental cycle as in Fig. 3.6 a) but after the evaporation we adiabatically load the atoms into an optical lattice. Before we take the picture, the lattice-beams are ramped down. Here, one can nicely see the first Brillouin zone. Its width in momentum space is $2\hbar k$. The schematic on the right (not true to scale) shows how the lattice is ramped up and down. In Fig. 3.6 b) it is crucial that the lattice is also ramped down adiabatically. This slow ramp ensures that the quasi-momentum does not change. We only observe the first Brillouin zone, indicating that the slow ramp down results in an occupation of the ground state of the optical lattice.

For our experiments it is crucial that we have only one molecule (two atoms) per lattice site. In the following I describe how we determine the occupation numbers for atoms experimentally. This question concerns the many-body state of the atoms, that is, whether we obtain a superfuid state or a Mott-insulator. We can answer this question with a slightly different sequence where the lattice is ramped up slowly (as shown in Fig. 3.6 b) but then switched off immediately. We show the BEC without an optical lattice in part c1). In c2) to c4) we increase the lattice depth slowly (200 ms) and vary the final value of the lattice depth. For shallow lattices we observe an interference pattern which we attribute to a coherent state with long-range phase coherence. c2) shows the superfluid state. Further increasing the lattice depth, we see a transition from a superfluid state to a Mott-insulator state in c4) where the phase coherence is lost completely. Analyzing the combination of the experiments corresponding to Figs. b) and c4), we conclude that a combination of deep lattices with an adiabatic rampdown results in a state where each lattice site is occupied with one atom in the ground state of the lattice.

The theory developed for atoms in optical lattices is valid for Feshbach molecules as well but one has to take twice the polarizability for molecules. This shows that Feshbach molecules can be described using their atomic properties. We thus have a trap depth of approximately $70\,E_r$ for our molecules.

### 3.6 Spin flip before Feshbach association

After loading the atoms into the optical lattice, they are in the $|f=1, m_f=-1\rangle$ state. Unfortunately, it turns out that the $|f=1, m_f=-1\rangle$ state does not show any Feshbach resonances. As the Feshbach resonance at 1007.4 G only appears for the $|f=1, m_f=1\rangle$ state [Marte 2002, Volz 2003] we have to transfer the molecules to this state. Therefore, we quickly change the direction of the magnetic bias field. The switching has to be done fast enough so that the total atomic angular momentum with its projection onto the space fixed axis cannot follow the magnetic field adiabatically. As it is difficult to change the direction of a current quickly, we ramp up a magnetic field through the offset coils (For the geometry of the coils see Fig. 3.3.) in 20 ms. Afterwards, we switch the field

3. Production of Feshbach molecules

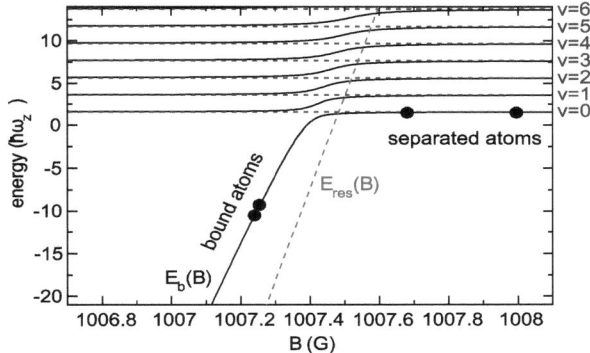

Figure 3.7: Feshbach resonance in an optical lattice. The horizontal dashed lines represent the 7 lowest trapped states of $^{87}$Rb atoms in the spin state $|f = 1, m_f = 1\rangle$ in the harmonic approximation. The trap frequency is set to $\omega_z = 2\pi \times 39\,\text{kHz}$. A molecular level (diagonal dashed line) crosses the atomic states at 1007.5 G. The coupling between the molecular and atomic states leads to an avoided crossing and thus connects the "separated atoms" state with the molecular Feshbach state. It also shifts the Feshbach resonance to 1007.4 G. Adapted from [Thalhammer 2007].

23

3. Production of Feshbach molecules

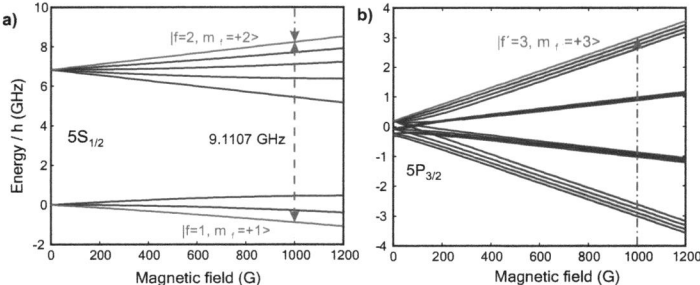

Figure 3.8: Hyperfine and magnetic structure of atomic $^{87}$Rb. For the purification scheme we drive a radio-frequency transition in the $5S_{1/2}$ state. Afterwards we excite to the $5P_{3/2}$ state to remove unbound atoms from the optical lattice. Adapted from [Winkler 2007].

through the offset coils off. The only remaining field is from the compensation cage[4]. Thus the direction of the magnetic field has changed which has the same effect as if the spin had flipped. The efficiency of this step is higher than 98% if the offset field from the compensation cage is set correctly. The efficiency can be checked with a Stern-Gerlach field which separates the different spin components before doing the absorption image. The efficiency of this step is crucial for the association of Feshbach molecules.

## 3.7 Ultracold Feshbach molecules in an optical lattice.

After the spin flip we start the Feshbach association. The principle of the Feshbach association in an optical lattice can be understood from Fig. 3.7. Here, we consider the magnetic field dependence of the molecular-state energy (bound atoms) and the energy dependence of the state consisting of "two separated atoms". The atoms oscillate in the optical lattice which forms an harmonic potential. Here, we only consider the zeroth vibrational level v = 0. This v should not be mixed up with the quantum number for molecular vibration introduced later. The two bare, that is non interacting, states show a crossing at a magnetic field $B \approx 1007.5$ G. The two states interact via the dominating exchange interaction and the weak spin-spin[5] interaction which lead to an avoided crossing. Thus, starting with two unbound atoms at $B \approx 1008$ G, lowering the magnetic field adiabatically, leads to the formation of a dimer. If we want to prepare a pure sample of Feshbach molecules we have to get rid of the huge number of atoms in singly occupied lattice sites. We can remove these unbound atoms with a combination of

---
[4]See remark on page 18.
[5]The Hamiltonians for effective spin-spin and spin-orbit interactions can not be distinguished. For brevity I omit the spin-orbit interaction although it is always present. For details see chapter 4.

# 3. Production of Feshbach molecules

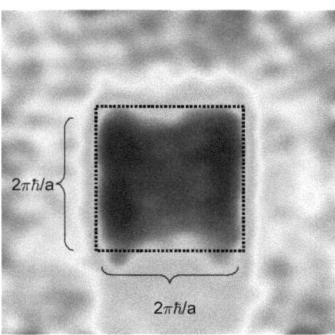

Figure 3.9: Feshbach molecules in the first brioullin zone after 13 ms time of flight. The horizontal and vertical dashed lines surround the first Brillouin zone. The bar structure results from repulsively bound pairs as described in Winkler et al. [Winkler 2006].

a radio frequency pulse and a resonant light pulse. Here we use an antenna to irradiate the atoms with 2 W of microwave radiation at 9.110 GHz. This electromagnetic field couples the $S_{1/2}$ $|f = 1, m_f = 1\rangle$ and $|f = 2, m_f = 2\rangle$ hyperfine states (Fig. 3.8). At the same time, we shine in light which is resonant with the $S_{1/2}$, $|f = 2, m_f = 2\rangle$ to $P_{3/2}$, $|f = 3, m_f = 3\rangle$ transition at a magnetic field of 0 G. At $B = 1000$ G this corresponds to a detuning of $\approx 1.4$ GHz. This light then "blows" away the remaining atoms, so all unbound atoms leave the lattice. A duration of 2 ms for these pulses is enough to get a pure sample of Feshbach molecules.

Fig. 3.9 shows an absorption picture of the dissociated molecules after the cleaning procedure. Unlike in Fig. 3.6 we now observe two vertical bars to the right and left-hand side of the first Brillouin zone. The structure consisting of two bars is typical for our molecules and indicates that we have created repulsively-bound pairs as described in [Winkler 2006]. In brief, these pairs show repulsive interaction and dissociate immediately in free space. However, in the lattice the energy which would be released in the dissociation corresponds to a forbidden area in the band structure. Thus, the repulsively bound pairs are stable. The bar which can be seen in Fig. 3.9 is a feature of proper Feshbach association because it does not appear when we have atoms in the optical lattice (Compare Fig. 3.6 b) ). We use this bar feature to check for proper Feshbach association when we perform our spectroscopy. For further details on repulsively bound pairs, I refer the reader to [Winkler 2006, Winkler 2007]

# 4 Theory of diatomic molecules

In the following chapter, I will show that some of the steps required to produce ultracold Feshbach molecules limit the number of accessible molecular states. Furthermore, I will review the basic concepts needed to understand the measured spectra presented in chapters 5 and 6. After some notation and definitions I start with the nonrelativistic Born-Oppenheimer and adiabatic approximation in section 4.2. In section 4.3, the results of a relativistic treatment are introduced, especially the electron and nuclear spin wavefunctions. In the first two sections, I will concentrate on the electronic term energies and the fine structure. All of the other interactions are explained in later chapters, always in the context of the measurements. Afterwards, Hund's coupling cases without nuclear spin are discussed (section 4.4). These cases give an intuitive understanding of the interactions which are present in the molecule using simple vector diagrams. The hyperfine structure sometimes dominates, but I will treat the Hund's cases without nuclear spin as a remainder. Hund's cases with nuclear spin are rarely found in literature [Townes/Schawlow 1955, Brown/Carrington 2003]. I will therefore review these cases in detail in the following section 4.5. The discussion of symmetry quantum numbers follows in section 4.6. Afterwards, I will apply the symmetries and Hund's cases to three important $^{87}$Rb$_2$ potentials: $X\,^1\Sigma_g^+$, $a\,^3\Sigma_u^+$, and $(1)\,^3\Sigma_g^+$ (section 4.7). The chapter concludes with the selection rules for bound-bound transitions in diatomic molecules in section 4.8.

## 4.1 Some notation: From Operators to exact and good quantum numbers

In the rest of the chapter we will use molecular wave-functions and molecular Hamilton operators which all involve angular momenta, for example $J$. In principle, $J$ can have the three different meanings

- an operator,
- a label in the basis function,
- and a good quantum number.

I will always indicate operators with bold face, for example the angular momentum operator $\mathbf{J}$. The basis functions are denoted by kets, in our example $|J\rangle$. If $J$ corresponds to the total angular momentum it is also a good quantum number. If it is not the total angular momentum $J$ is in most cases only a label which does not yield a good quantum number. The latter just means that the eigenstates of the Hamiltonian are superpositions of different $|J\rangle$ (See [Lefebvre-Brion/Field 2004], page 71.). It has to be clear from the

# 4. Theory of diatomic molecules

| Type of angular momentum | Operator | Quantum numbers | |
|---|---|---|---|
| | | Total | Projection |
| Nuclear rotation..................... | **R** | $R$ | ... |
| Electronic orbital.................... | **L** | $L$ | $\Lambda$ |
| Total without spins................... | $\mathbf{N} = \mathbf{R} + \mathbf{L}$ | $N$ | $\Lambda$ |
| Electronic spin ...................... | **S** | $S$ | $\Sigma$ |
| Orbital with electron spin............ | $\mathbf{J}_a = \mathbf{L} + \mathbf{S}$ | $J_a$ | $\Omega$ |
| Total with electron spin ............. | $\mathbf{J} = \mathbf{R} + \mathbf{L} + \mathbf{S}$ | $J$ | $\Omega$ |
| Nuclear spin......................... | **I** | $I$ | $\Omega_I$ |
| Total with electron and nuclear....... | $\mathbf{F} = \mathbf{R} + \mathbf{L} + \mathbf{S} + \mathbf{I}$ | $F$ | $\Omega_F$ |

Table 4.1: Important angular momenta in Hund's coupling cases. The projection always refers to the molecule-fixed (internuclear) axis. Space-fixed projections are always labelled with subscript $z$, for example $\mathbf{L}_z$. The molecule-fixed projection of **R** is not identified as it is zero. $R$ should not be confused with the internuclear distance $R$ introduced later.

context whether $J$ is a good quantum number or just a label in this function. In the nomenclature I will follow Nikitin [Nikitin 1994], who additionally distinguishes between exact quantum numbers and good quantum numbers. Exact quantum numbers are always good quantum numbers, which means that they do not depend on the angular momentum coupling scheme. Examples for exact quantum numbers are the total energy $E$ and the total angular momentum. Good quantum numbers depend on the coupling scheme, that is on the Hund's case. For example $\Lambda$ (see Table 4.1) is not an exact quantum number but in Hund's case a) it is a good quantum number. On the other hand it turns out that $\Lambda$ is a bad quantum number in Hund's case c). This will be explained in detail in section 4.4. The difference between good and exact quantum numbers is going to be important when we anlayse selection rules between different Hund's cases.

## 4.2 Nonrelativistic Born-Oppenheimer and adiabatic approximation

In this section I follow the reasoning of [Bunker 1968] but I skip some details and some symbols are changed so that they are consistent with the nomenclature in the book of Lefebvre-Brion and Field [Lefebvre-Brion/Field 2004].
The configuration of our $^{87}$Rb atoms in the electronic ground state is (Kr)5s, where the closed Krypton shell is denoted by (Kr). Its electronic term in the ground state is therefore $5S_{1/2}$. The first excited state corresponds to a fine-structure doublet $5P_{1/2} + 5P_{3/2}$, where the $5P_{1/2}$ state has less energy. The ground state of a $^{87}$Rb$_2$ molecule connects to the $5S_{1/2} + 5S_{1/2}$ two-atom asymptote. This can be seen in Fig. 4.1 for $^{87}$Rb$_2$ [Lozeille 2006]. For large nuclear separation, the atomic configuration is retrieved. If the atoms get close to each other, bound molecular states form. This can be seen from the

# 4. Theory of diatomic molecules

minima in the potential curves. It is crucial to have an overview over the wave-functions and the Hamiltonian with its approximations to understand the symmetries. This is reviewed briefly in this section for the nonrelativistic case.

The Hamiltonian for a diatomic $^{87}_{37}\text{Rb}_2$ molecule consists of 2 nuclei and $2\times 37$ electrons and the corresponding Schrödinger equation is given by

$$\mathbf{H} \cdot \Psi^T = E \cdot \Psi^T, \quad \text{with}$$
$$\mathbf{H} = \mathbf{H}^{nrel} + \mathbf{H}^{rel} + \mathbf{H}^{Lamb}, \quad \text{and} \quad (4.1)$$
$$\mathbf{H}^{nrel}(R,\theta,\phi,\mathbf{r}) = \mathbf{T}^N(R,\theta,\phi) + \mathbf{T}^e(\mathbf{r}) + \mathbf{V}(\mathbf{r},R).$$

Here we have the nonrelativistic operator $\mathbf{H}^{nrel}$ and the relativistic Hamiltonian $\mathbf{H}^{rel}$. The latter part includes, for example, spin-orbit and spin-spin interactions. $\mathbf{H}^{Lamb}$ includes interactions with the electromagnetic field and can be neglected. $\mathbf{H}^{nrel}$ involves the internuclear separation $R$. The Euler angles $\theta$ and $\phi$ specify the orientation of the molecule, and $\mathbf{r} = (\mathbf{r}_1, \cdots, \mathbf{r}_{2\times 37})$ is the position vector including all $2\times 37$ electrons. $\mathbf{T}^N(R,\theta,\phi)$ ($\mathbf{T}^e(\mathbf{r})$) is the operator for the kinetic energy of the nuclei (electrons), and $\mathbf{V}(\mathbf{r},R)$ contains the potentials, including electron-electron, nuclei-nuclei, and electron-nuclei Coulomb interactions.

In a first step, we only consider the nonrelativistic electronic part, where we assume that the electrons follow the nuclei adiabatically. Separation of the electronic and nuclear motion in Eq. (4.1) leads to

$$\mathbf{H}^e \cdot \Psi_n^e(\mathbf{r};R) \equiv (\mathbf{T}^e(\mathbf{r}) + \mathbf{V}(\mathbf{r},R)) \cdot \Psi_n^e(\mathbf{r},R) = E_n^0(R) \cdot \Psi_n^e(\mathbf{r};R), \quad (4.2)$$

where we have used the product ansatz $\Psi_n^T = \Psi_n^e \cdot \Psi_n^{rv}$ to separate the electronic ($\Psi_n^e$) from the ro-vibronic motion ($\Psi_n^{rv}$). In this equation, we treat $R$ as a parameter and not as variable. The index $n$ then labels the different eigenstates with corresponding $R$ dependent eigenvalues $E_n^0(R)$. As in the case of atoms, we can build the total electronic angular momentum $\mathbf{L}$. However the operator $\mathbf{H}^e$ breaks the rotational symmetry. Thus, only the molecule fixed $Z$-component $L_Z$ commutes with $\mathbf{H}^e$ because this direction is an axis of symmetry. Only the projection $\Lambda$ of $\mathbf{L}$ onto the internuclear axis has well defined values [Kronig 1930] and we obtain wave-functions of the form $\Psi_{n,\Lambda}^e(\mathbf{r};R)$. The corresponding eigenenergies are denoted by $E_{n,\Lambda}^0(R)$ and are called potential curves. The various quantum numbers which we need in this chapter are summarized in Table 4.1.

**Remark.** The wave-functions $\Psi_{n,\Lambda}^e(\mathbf{r};R)$ and eigenenergies $E_{n,\Lambda}^0(R)$ are the result of *ab-initio* calculations and it is useful to write $\Psi_{n,\Lambda}^e(\mathbf{r};R) = \langle \mathbf{r};R \mid n\Lambda \rangle$. This simplifies calculations if we work with effective Hamiltonians where matrix elements are given as $\langle n'\Lambda' \mid H_{eff} \mid n\Lambda \rangle$. The potential curves in Fig. 4.1 are the $R$ dependent expectation values for the various states $\mid n\Lambda \rangle$. Here, $\Sigma$ states have $\Lambda = 0$, $\Pi$ states have $\Lambda = 1$, and so forth. The other labels will be explained later as they are a result of relativistic effects (spin of the electron).

# 4. Theory of diatomic molecules

Next, we consider the full nonrelativistic problem, including molecular vibration and rotation. We decompose the total molecular wave-function in the form

$$\Psi(\mathbf{r}, R) = \sum_n \Psi^e_{n,\Lambda}(\mathbf{r}; R) \cdot \Psi^{rv}_n(R, \theta, \phi) = \sum_n \Psi^e_{n,\Lambda}(\mathbf{r}; R) \cdot \Psi^{rot}_n(\theta, \phi) \cdot \Psi^{vib}_n(R), \quad (4.3)$$

where the second equation holds if the rotational and vibrational motions are separable. A wave-function of the form $\Psi^e_{n,\Lambda}(\mathbf{r}; R) \cdot \Psi^{rv}_n(R, \theta, \phi)$ is called a Born-Oppenheimer product function. Inserting this ansatz into the nonrelativistic Schrödinger equation $\mathbf{H}^{nrel}\Psi^T = E\Psi^T$ leads to an infinite system of coupled differential equations of the form

$$\left(E^0_n(R) - E\right) \cdot \Psi^{rv}_n(R, \theta, \phi) + \sum_{n'} C^0_{nn'} \cdot \Psi^{rv}_{n'}(R, \theta, \phi) = 0$$

$$C^0_{nn'} = \int \cdots \int \Psi^e_{n,\Lambda}(\mathbf{r}; R)^* \cdot \left[-\mathbf{T}^N(R, \theta, \phi)\right] \cdot \Psi^e_{n',\Lambda'}(\mathbf{r}; R) \, d\mathbf{r}. \quad (4.4)$$

In the **Born-Oppenheimer approximation** we assume the nuclei to have infinite mass. Furthermore we neglect the $R$ dependence of $\Psi^e_{n,\Lambda}$. This causes all interactions between different electronic states to vanish, that is $C^0_{nn'} = 0$ if $n \neq n'$ and $C^0_{nn} = -\mathbf{T}^N(R, \theta, \phi)$. The remaining equation is the simplest one, but neglects any interaction between electrons and nuclei. As eigenfunctions for the rotational part of $\mathbf{T}^N$ we can take the Wigner rotation matrices $\mathbf{D}^{(N)}_{\Lambda m_N}(\pi/2, \theta, \phi) \sim \langle \pi/2, \theta, \phi \mid NM\Omega \rangle$ (See for example [Edmonds 1960].). Here, the total angular momentum $\mathbf{N}$ is the sum of pure molecular rotation $\mathbf{R}$ and total orbital angular momentum $\mathbf{L}$. The angle of $\pi/2$ arises because a diatomic molecule has only two principal axes of rotation. The third angle then only fixes an overall phase. In the Born-Oppenheimer limit we therefore end up with

$$\left(-\mathbf{T}^N(R, \theta, \phi) + E^0_{n,\Lambda}(R) - E\right) \cdot \Psi^{rv}_n(R, \theta, \phi) = 0. \quad (4.5)$$

The **Adiabatic approximation** is used when one includes the adiabatic interaction between nuclei and electrons. Again only terms of the form $C^0_{nn}$ are included in the non relativistic equation (4.4) and lead apart from the $-\mathbf{T}^N(R, \theta, \phi)$ term to a modified Born-Oppenheimer potential

$$E^{ad}_n(R) = E^0_n(R) + A^0_{nn}(R). \quad (4.6)$$

The $A^0_{nn}(R)$ term arises from the $R$-dependence of the electronic wave-function $\Psi^e_{n,\Lambda}(r; R)$. (For details see the excellent article by [Bunker 1968].) The final equation is the same as in the Born-Oppenheimer approximation except that we replace the effective potential $E^0_n(R)$ by $E^{ad}_n(R)$ according to Eq. (4.6). Adiabatic potentials show sometimes an avoided crossing like behavior.

## 4.3 Treatment with electronic and nuclear spins

The correct treatment of relativistic effects requires some effort but the steps are similar to those in the nonrelativistic case. Relativistic effects include spin-spin, spin-orbit, and spin-rotation interactions. I will only give the important results. (See

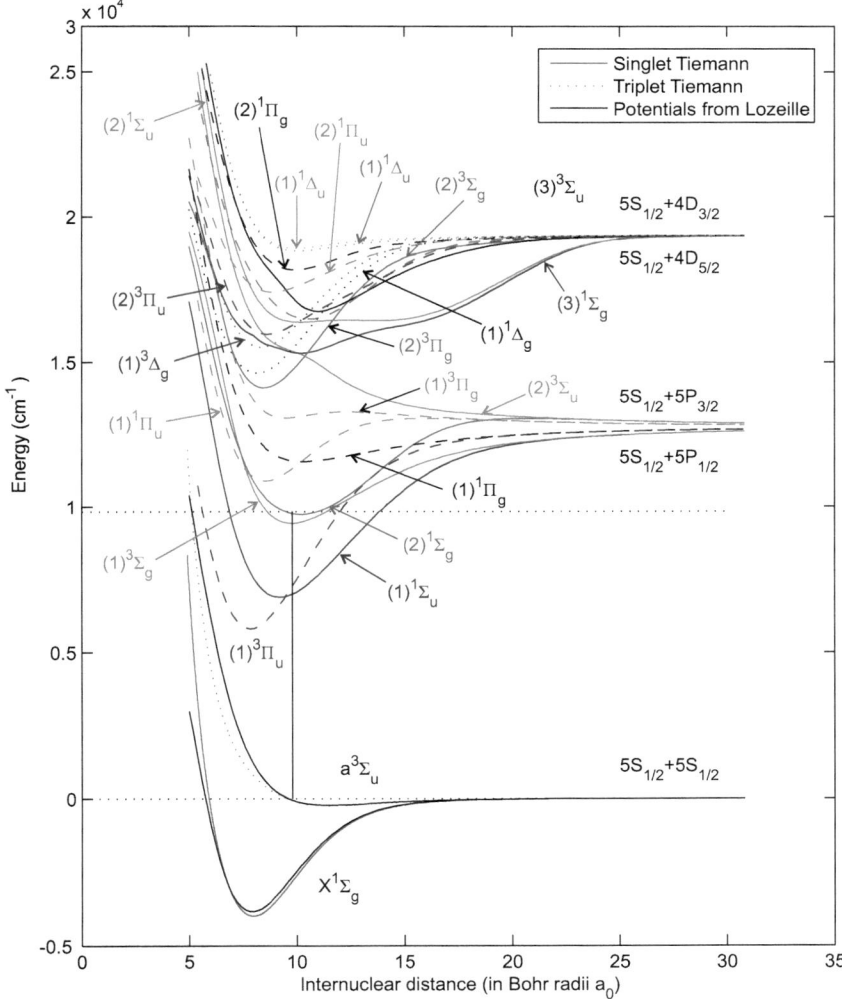

Figure 4.1: Potential scheme of the $^{87}\text{Rb}_2$ molecule from [Lozeille 2006] with the two optimized ground-state potentials from Tiemann (red). The solid (dotted) red curve corresponds to the optimized singlet (triplet) potential from E. Tiemann [Strauss 2010] (For details see also chapter 6.). All other potentials are from Lozeille *et al.* . Solid curves correspond to $\Sigma$ states, dashed (dotted) curves correspond to $\Pi$ ($\Delta$) potentials. The vertical bar corresponds to the energy of our one-photon laser (see chapter 6).

## 4. Theory of diatomic molecules

[Kronig 1930, Bunker 1968, Brown/Carrington 2003] for a relativistic treatment of the fine structure.) Compared to the nonrelativistic Hamiltonian, it is clear that we start with new eigenfunctions of the form

$$\Psi_n^T = \Psi_n^e \cdot \chi_n \cdot \beta_n \cdot \Psi_n^{rot} \cdot \Psi_n^{vib} = \Psi_n \cdot \beta_n. \quad (4.7)$$

Here we introduce the new spin functions $\chi_n$ and $\beta_n$ for electronic and nuclear spins, respectively. In contrast to the nonrelativistic case, where we only had the quantum numbers $n$ and $\Lambda$, we must now consider the various spins. These functions will depend on the interactions present in the molecule. This can be understood as follows: The total electronic spin $\mathbf{S}$ has a magnetic moment which can interact with the magnetic field due to the total electronic orbital angular momentum $\mathbf{L}$ (spin-orbit or $\mathbf{L} \cdot \mathbf{S}$ interaction). According to the strength of this interaction, it is either included in $\mathbf{H}^e$ or excluded. The case of weak spin-orbit coupling is referred to as Hund's case a) and the electronic eigenfunctions become $\mid n\Lambda S\Sigma\rangle$[1]. The opposite case is called Hund's case c). We then have to replace $\Lambda$ by $\Omega$, the molecule-fixed projection of the total angular momentum $\mathbf{J} = \mathbf{N} + \mathbf{S}$. In Hund's case a) this connects of course to $\Omega = \Lambda + \Sigma$. Here, the basis consists only of $\mid n\Omega\rangle$, the case with the fewest good quantum numbers. (See section 4.5.2 and Table 4.2.)

### 4.4 Hund's cases without nuclear spin

At the advent of quantum mechanics, Hund gave an explanation for the term schemes [Hund 1927] and the observed structure in diatomic molecules [Hund 1927a]. In the latter paper, he suggested a classification based on a classical picture of angular momenta. These limiting cases are an idealized interaction scheme. One can find unitary transformations between the different basis sets, and thus the eigenenergies do not depend on the basis as expected. A nice summary of these ideas, along with vector diagrams, can be found in [Hund 1933]. In the following, I will show the "classical" Hund's cases as a reminder (see Fig. 4.2 and Table 4.2).

In Hund's case a), the total electronic orbital angular momentum $\mathbf{L}$ and the total electronic spin $\mathbf{S}$ are coupled separately to the internuclear axis. Thus, $\Lambda$ and $\Sigma$ are the projection quantum numbers. The total angular momentum $\mathbf{J}$ is composed of $\mathbf{R}$, $\mathbf{L}$, and $\mathbf{S}$. Here, the good quantum numbers in the rotating molecule are $\Lambda$, $\Sigma$, $\Omega = \Lambda + \Sigma$, $J$, and $m_J$. In most cases $S$ is also a good quantum number. This is explained nicely in the appendix of [Zare 1973]

In case c), the total electronic orbital angular momentum and the total electronic spin are strongly coupled. $\Lambda$ and $\Sigma$ are now bad quantum numbers. However, the projection $\Omega$, which equals $\Lambda + \Sigma$ in case a), remains good. Strictly speaking, this is the only good quantum number. Thus, Hund's case c) is the case with the fewest good quantum

---
[1] These functions build the Hund's case a) basis. Whether these labels $n$, $\Lambda$, ... are good quantum numbers or not has to be calculated *ab initio*.

4. Theory of diatomic molecules

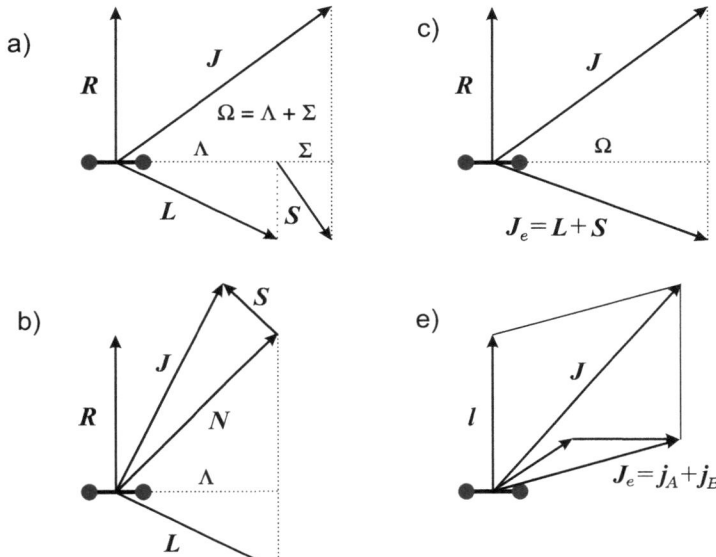

Figure 4.2: Hund's cases without nuclear spin. The pure mechanical rotation of the molecule is denoted by **R** except for case e) where I denote it **l**. An overview of all of the other quantum numbers can be found in Table 4.1.

# 4. Theory of diatomic molecules

| Coupling case | Good quantum numbers | $H^{el}$ | $H^{SO}$ | $H^{rot}$ | Nomenclature |
|---|---|---|---|---|---|
| (a) | $\eta, \Lambda, S, \Sigma, J, (\Omega = \Lambda + \Sigma)$ | s | i | w | $^{2S+1}\Lambda_{\Omega,\iota_e}(\Lambda \neq 0)$ |
| (b) | $\eta, \Lambda, N, S, J$ | s | w | i | $^{2S+1}\Lambda_{\iota_e}(\Lambda \neq 0), {}^{2S+1}\Sigma^{\sigma_{el}}_{\iota_e}(\Lambda = 0)$ |
| (c) | $\eta, \alpha, J, \Omega$ | i | s | w | $\Omega_{\iota_e}(\Omega \neq 0), 0^{\sigma_{el}}_{\iota_e}(\Omega = 0)$ |
| (d) | $\eta, L, R, N, S, J$ | i | w | s | not established |
| (e) | $\eta, J_a, J$ | w | i | s | not established |

Table 4.2: Overview over Hund's coupling cases without nuclear spin. The second column gives the good quantum numbers and the third column shows the electronic, spin-orbit, and rotational interactions with their strength. "w" means weak, "i" intermediate, and "s" stands for strong. In the last column $\iota_e$ is the quantum number corresponding to the inversion of electron coordinates only. $\sigma_{el} = \pm 1$ is the eigenvalue for the reflection with respect to a plane containing the internuclear axis (For details see section 4.6.2 at 43.). For details on the symmetry operations see section 4.6.2. All of the other quantum numbers are defined in Table 4.1. For Hund's cases d) and e) no common nomenclature exists. The good quantum numbers given here should not be mixed up with those of Hougen [Hougen 1970] because he always refers to the non-rotating molecule. Adapted from [Nikitin 1994].

numbers. However the total angular momentum $J$ is the same as in case a) and the rotational energy ladder goes as $J(J+1)$. If one needs a Hund's case c) basis, an additional quantum numbers $\alpha$ has to be introduced. This quantum numbers then assures that a unique transformation to the other Hund's cases exists.

In case b), the strong rotational interaction decouples the spin from the axis but the total orbital angular momentum is still coupled to the internuclear axis. The rotational energy ladder is given by $N(N+1)$ where $\mathbf{N} = \mathbf{R} + \mathbf{L}$. The good quantum numbers in this case are $\Lambda$, $N$, $S$, $J$, and $m_J$. Case b) is normally important for $\Sigma$-states where the projection $\Lambda$ of the total orbital angular momentum onto the internuclear axis is 0, except when there is a strong spin-spin interaction.

Hund's case e) is the analogue to the jj-coupling in atoms and will become important for our Feshbach molecules. In this case, l also includes the angular momenta of the electrons which is in contrast to the $\mathbf{R}$ vector. However, the difference between the l and $\mathbf{R}$ vectors is negligible due to to the small mass of the electrons (compared to the nuclei). As the hyperfine structure is dominant for these weakly bound molecules, it will be considered in detail in the next section.

In Hund's case d) $\mathbf{L}$ is decoupled from the molecular axis, which happens for example in Rydberg states. As this coupling scheme is not important for the $^{87}$Rb$_2$ molecule, we will not discuss it further.

The cases are summarized in Table 4.2. The quantum numbers given in the second column hold for the rotating molecule. (Compare Hougen [Hougen 1970] for the nonrotating molecule.) If we include the symmetries from section 4.6, $\Lambda$, $\Sigma$, and $\Omega$ are always positive. $\eta$ represents quantum numbers which specify the electronic state. It includes $n$, and symmetry quantum numbers like $\sigma_{el}$ and $w$. For a transformation to case c) in the above scheme it is necessary to introduce a new quantum number $\alpha$. As already mentioned, it would otherwise not be possible to transform the state vector for example to a Hund's case a) state vector. Most molecules belong to intermediate coupling schemes and the quantum numbers in a state vector become bad. This will be discussed in more detail with the experimental results.

## 4.5 Hund's coupling cases including nuclear spin

Hund's coupling cases including nuclear spins are more complicated because of additional interactions. The basic idea behind the nomenclature is similar to that behind the classical Hund's cases a) and b). In case a) the electronic spin is coupled to the internuclear axis with molecule-fixed projection $\Sigma$, whereas in case b) it is decoupled. To avoid confusion with the electronic spin, one adds indices $\alpha$ or $\beta$ depending on whether the nuclear spin is coupled or decoupled from the internuclear axis, respectively. I will discuss these schemes in detail for case $a_\alpha$) and $a_\beta$). For all of the other Hund's cases, the reasoning is similar. Therefore, I will only consider the situation when the nuclear spin is coupled to the internuclear axis in all of the other cases.

When looking at the spin functions, one might think that there are a large number of spins which contribute to the coupling. Luckily, this is not really the case because most of the spins are absorbed in inner (closed) shells and thus do not contribute. In fact, it turns out that the molecular electronic configuration of $^{87}\text{Rb}_2$ is given by the two valence electrons. In principle, we have two spins with corresponding bases $\mid s_1 m_{s_1}\rangle$ and $\mid s_2 m_{s_2}\rangle$. On the other hand, we could also work in a coupled scheme $\mid (s_1, s_2) S m_S\rangle$. As it is clear that the physics does not depend on the basis in use, we will always consider the molecular basis with quantum number $S$ in the following and omit the two constants $s_1$ and $s_2$. The same holds for the nuclear spin $I$.

### 4.5.1 Case a) with basis functions

As the nuclear magnetic moment is much weaker than the electronic moment there are basically two possibilities in case a). When the nuclear spin $I$ is coupled to the axis we have case $a_\alpha$) as can be seen in Fig. 4.3. Here we have the most good quantum numbers including $\Lambda$, $S$, $\Sigma$, $\Omega = \Lambda + \Sigma$, $I$, $\Omega_I$, and $F$ at zero magnetic field. Furthermore, the space- and molecule-fixed projections of the total angular momentum $M_F$ and $\Omega_F = \Omega + \Omega_I$ are good. At zero magnetic field the total angular momentum $F$ is also a good quantum number.

# 4. Theory of diatomic molecules

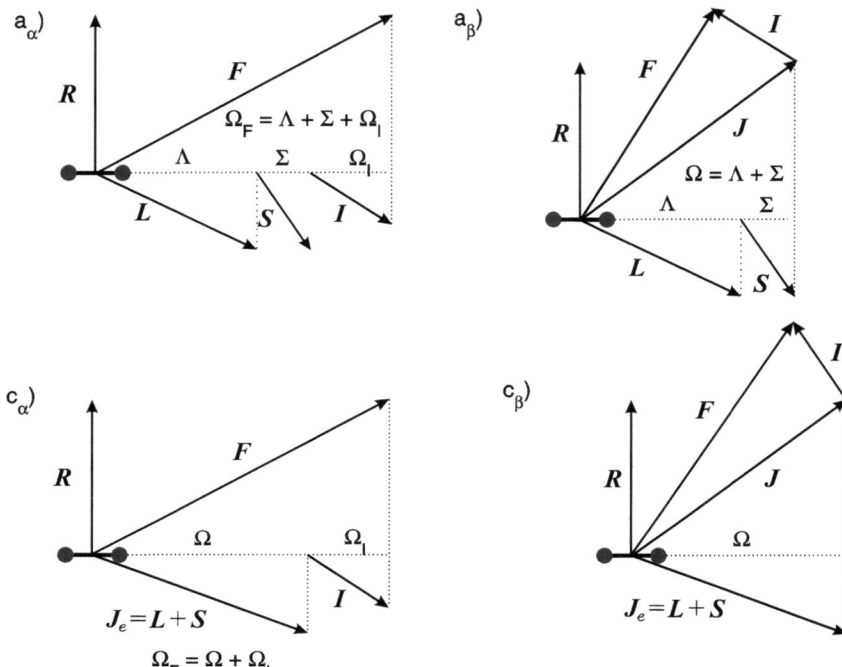

Figure 4.3: Hund's cases a) and c) including nuclear spin. The quantum numbers are defined in Table 4.1.

As Hund's coupling cases depend on the strength of different interactions, the wave-functions depend on the choice of a coupling scheme and especially on the order of coupling. This means that they are not the bare product functions of Eq. 4.7 and for example in case $a_\alpha$) the corresponding wave-functions are of the form

$$\Psi^e = \Psi^e_q(\mathbf{r}; R)$$
$$\Psi^{vib} = \Psi^{vib}_{q,v}(R)$$
$$\Psi^{rot} = \Psi^{rot}_{F,m_F,\Omega_F}(\omega)$$
$$\chi = \chi_{S,\Sigma}(\omega)$$
$$\beta = \beta_{F,\Omega_F}(\omega).$$

Here, we use the abbreviations $q = (n, \Lambda, S, \Sigma, \Omega)$ and $\omega = (\pi/2, \theta, \phi)$ for an ideal diatomic molecule which cannot rotate around the internuclear axis. The angle is in accordance with the phase choice of Lefebvre-Brion/Field [Lefebvre-Brion/Field 2004]. It should be clear that the vibrational wave-function is scalar and thus $q$ just labels these functions. If we write $\chi_{S,\Sigma}$ we mean implicitly

$$\chi_{S,\Sigma}(\omega) = \sum_{m_S} \sqrt{\frac{2S+1}{4\pi}} D^{(S)}_{m_S,\Sigma}(\omega),$$

where we use the rotation matrix $D^{(S)}_{m_S,\Sigma}(\omega)$ for a diatomic molecule. The coupling scheme above means that we assume the spin-orbit coupling to be weak and thus the electronic spin $\mathbf{S}$ with projection $\Sigma$ couples first to the axis because of the interaction between $\mathbf{N} = \mathbf{R} + \mathbf{L}$ and $\mathbf{S}$. Afterwards, the nuclear spin also couples to the internuclear axis because of the interaction with $\mathbf{J} = \mathbf{R} + \mathbf{L} + \mathbf{S}$ to form a total angular momentum $\mathbf{F} = \mathbf{J} + \mathbf{I}$ with projection $\Omega_F$. The molecular rotation is given by

$$\Psi^{rot}_{F,m_F,\Omega_F}(\omega) = \langle \omega \mid Fm_F\Omega_F \rangle = \sqrt{\frac{2F+1}{4\pi}} D^{(F)}_{m_F,\Omega_F}(\omega)$$

with corresponding energy eigenvalues $\hbar^2 F(F+1)$. The equation involving the rotation matrices uses standard angular-momentum theory and can be found in the literature, for example in [Edmonds 1960, Brink/Satchler 1971, Rose 1957] and [Lindner 1984].

Case $a_\beta$) is the same except for the decoupling of the nuclear spin which means that $\Omega_I$ becomes bad. The wave-functions would look like

$$\Psi^e = \Psi^e_q(\mathbf{r}; R)$$
$$\Psi^{vib} = \Psi^{vib}_{q,v}(R)$$
$$\Psi^{rot} = \Psi^{rot}_{F,m_F,\Omega_F}(\omega)$$
$$\chi = \chi_{S,\Sigma}(\omega)$$
$$\beta = \begin{cases} \beta_{J,m_J,I,m_I} & I \text{ and } J \text{ uncoupled} \\ \beta_{(JI)F,m_F} & \text{coupled basis.} \end{cases}$$

Here, the coupling scheme up to $J$ is similar to that in case $a_\alpha$). There are the limiting possibilities of strong (fully coupled) or weak hyperfine coupling (uncoupled). It should be clear that it is impossible to distinguish the two hyperfine coupling cases when looking at Fig. 4.3. As in the classical Hund's, cases one can distinguish between the cases $a_\alpha$) (rotational energy ladder goes as $F(F+1)$) and $a_\beta$) (rotational energy ladder goes as $J(J+1)$) by looking at the observed spectra (compare Nikitin *et al.* for the case without nuclear spin [Nikitin 1994]). Normally, the electronic and nuclear spin functions as well as the rotational wave-functions are included in $\Psi^e$ and we obtain molecular basis functions of the form

$$\Psi^e_q(\mathbf{r};R) \cdot \chi_{S,\Sigma}(\omega) \cdot \beta_{F,\Omega_F}(\omega) \cdot \Psi^{rot}_{F,m_F,\Omega_F}(\omega) = \langle \mathbf{r}; R, \omega \mid n\Lambda \ S\Sigma \ I\Omega_I \ Fm_F \rangle \quad (4.8)$$

for case $a_\alpha$). If one wants to picture the rotation of the molecule, one has to go to the decoupled basis because spin wave-functions are hard to picture as they turn the wave-function into its negative under a rotation of $2\pi$. The procedure is the same for Hund's case $a_\beta$) and I will skip the lengthy formulas.

Thus, we arrive at basis vectors $|\ n\Lambda\ S\Sigma\ I\Omega_I\ Fm_F\rangle$, $|\ n\Lambda\ S\Sigma\ (JI)Fm_F\rangle$, and $|\ n\Lambda\ S\Sigma\ Jm_J\ Im_I\rangle$ in the different couplings schemes, which we can use for example to implement effective Hamiltonians [Lefebvre-Brion/Field 2004, Brown/Carrington 2003]. An appropriate choice of basis for a special problem will lead to small off diagonal matrix elements and speed up the calculations. In other words, if the eigenvectors are close to one basis one can learn something about the strengths of the different interactions.

### 4.5.2 Case c) with basis functions

The reasoning is similar for the cases $c_\alpha$) and $c_\beta$) but $\Lambda$ and $\Sigma$ are not good anymore. Therefore $\Lambda$ and $\Sigma$ are missing in Fig. 4.3). The physical reason for only $\Omega$ remaining good is a strong spin-orbit interaction which couples states with different $\Lambda$ and $\Sigma$. In a pure Hund's case c) state the only good quantum number is $\Omega$. If one studies for example the molecular states which connect to the $^2P_{1/2} +^2 P_{3/2}$ asymptote, we find 8 Hund's case a) states but 16 Hund's case c) states. (See for example [Wang 1997] and the review article by [Jones 2006].) This reflects the fact that we have fewer quantum numbers in case c). But how can we then transform a state given in this basis to a different representation, say Hund's case a)? As mentioned previously, we have to extend the state vector using additional quantum numbers, so that there is a unitary transformation between the state vectors of cases a) and c). These numbers represent the different coupling schemes between $\mathbf{R}$, $\mathbf{L}$, $\mathbf{S}$, and $\mathbf{I}$, which lead to one Hund's case c) state. A basis vector thus looks like

$$\Psi^e = \Psi^e_{n,\alpha,\Omega}(\mathbf{r};R)$$
$$\Psi^{vib} = \Psi^{vib}_{n,\alpha,\Omega,v}(R)$$
$$\Psi^{rot} = \Psi^{rot}_{F,m_F,\Omega_F}(\omega)$$
$$\chi = \chi_{S,\Sigma}$$
$$\beta = \beta_{F,\Omega_F}(\omega),$$

when the electronic wave-functions are evaluated in a Hund's case c) basis. Here we use the same notation as for the various a) cases. In a pure Hund's case $c_\alpha$) molecule, the only good quantum numbers in the rotating molecule are $\Omega$ and $\Omega_I$. I omit $c_\beta$) as the ideas are the same as for Hund's cases $a_\alpha$) and $a_\beta$). In our shortened notation our state vectors are $|n, \alpha, \Omega, I, \Omega_I, F, \Omega_F\rangle$, and the case $c_\beta$) vectors $|n, \alpha, \Omega, (JI)F, M_F\rangle$, and $|n, \alpha, \Omega, J, M_J, I, M_I\rangle$.

### 4.5.3 Case b) with basis functions

In Hund's coupling scheme b) we encounter three different situations according to the coupling order of $N$, $S$, and $I$. When the electronic spin couples to the rotation $N$ the scheme is called case b$_{\beta S}$). In the other two cases the nuclear spin couples first to $N$ or there is a strong coupling between $S$ and $I$. Both result in an intermediate vector $f$ as shown in Fig. 4.4. As the spins are not coupled to the axis, it is not useful to refer to Hund's case b) potentials [Lefebvre-Brion/Field 2004]. Therefore, I decided to just give the electronic wave-functions in the case a) basis. Here the case b$_\alpha$) wave-functions look like

$$\Psi^e = \Psi^e_q(\mathbf{r}; R)$$
$$\Psi^{vib} = \Psi^{vib}_{q,v}(R)$$
$$\Psi^{rot} = \Psi^{rot}_{F, m_F, \Omega_F}(\omega)$$
$$\chi = \chi_{S, m_S}$$
$$\beta = \beta_{I, \Omega_I}(\omega),$$

where we use the abbreviation $q = (n, \Lambda, S, \Sigma, \Omega)$. In this case the good quantum numbers in the rotating molecule include $N, \Lambda, S, I,$ and $\Omega_I$. Thus, the state vectors are $|nN\Lambda\ S\ I\Omega_I\ F\Omega_F\rangle$. In the various b$_\beta$) cases we obtain $|nN\Lambda\ S\ JM_J\ IM_I\rangle$, $|nN\Lambda\ S\ (JI)FM_F\rangle$ and $|nN\Lambda\ (SI)f, FM_F\rangle$.

### 4.5.4 Case e) with basis functions

In case e) the total atomic angular momentum including the nuclear spins $f_a$ ($f_b$) of atoms $a$ ($b$) couple to form $f$. Afterwards, the pure molecular rotation couples in to form the total angular momentum $F$. It is clear that the nuclear spin cannot be coupled to the axis in this coupling case. Even when one includes the nuclear spin, there is only one case e). In the coupled basis this can be written as

$$\Psi^e \cdot \chi \cdot \beta = \Psi_q(\mathbf{r}; R)$$
$$\Psi^{vib} = \Psi^{vib}_{q,v}(R)$$
$$\Psi^{rot} = \Psi^{rot}_{F, M_F, \Omega_F}(\omega)$$

where $q = (n, f_a, f_b, f, l, F, M_F)$. This gives rise to the fully coupled basis $|\ n\ (f_a f_b) f l F M_F\rangle$. Again, we can for example choose a partially coupled basis $|\ n\ (f_a f_b) f M_f\ l M_l\rangle$ or the uncoupled basis $|\ n\ f_a M_{f_a}\ f_b M_{f_b}\ l M_l\rangle$. As already mentioned our Feshbach molecules exhibit strong case e) structure. This is due to the fact that these molecules are close to dissociation threshold and can be characterized by their atomic properties.

## 4.6 Symmetries

After we had a look at the Hamiltonian and the classification of its energy scales, we focus on the symmetries of diatomic molecules. As a first step, we consider exchange

# 4. Theory of diatomic molecules

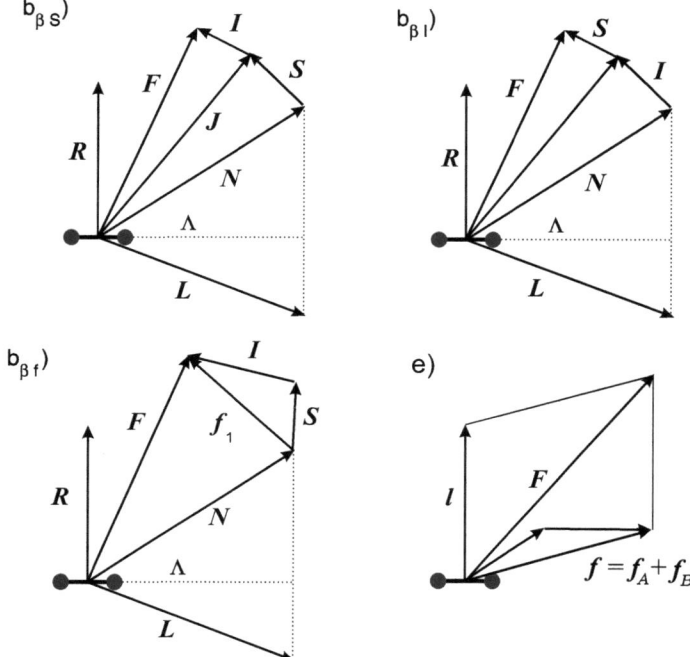

Figure 4.4: Hund's cases b) and e) with nuclear spin. The pure mechanical rotation of the molecule is denoted by **R** except for case e) where it is denoted by **l**. The quantum numbers are defined in Table 4.1.

symmetries in order to find the accessible Feshbach states. In these calculations it turns out that the quantum numbers of our Feshbach states follow simple sum rules of the form $S + I = 0, 2, ....$ In the next step we consider the different restrictions with respect to symmetry when a molecule forms. This will explain the remaining quantum numbers in Table 4.2. These results will be applied to the relevant states in the next section.

### 4.6.1 Exchange symmetries and Feshbach molecules

It is well known that a many-body system consisting of bosons possesses a symmetric wave-function with respect to the exchange of two particles. For fermions the function is antisymmetric [Sakurai 1995]. This has consequences not only for our initial Feshbach state but for all molecular states. In the analysis it is crucial that we are dealing with $\Sigma$-states, that is a state with $\Lambda = 0$ because only in this special case is there no influence of $\mathbf{L}$.

**Atomic basis.** As it is not a great effort to treat the bosonic case and the fermionic case at once, I will consider the symmetrization operator

$$1 + (-1)^p P_{12}. \tag{4.9}$$

Here $p = 0$ ($p = 1$) for the bosonic (fermionic) case and $P_{12}$ is an operator which exchanges atoms $a$ and $b$. In the center of mass system the spatial part of this operator just equals parity, that is all coordinates are replaced with their negatives. When we apply the operator $P_{12}$ to the spherical harmonics $Y_{lm} = \langle \phi, \theta \mid lm_l \rangle$ we obtain

$$P_{12} \mid lm_l \rangle = (-1)^l \mid lm_l \rangle. \tag{4.10}$$

In the atomic basis this leads to

$$(-1)^p P_{12} \mid f_a m_{f_a}, f_b m_{f_b}, lm_l \rangle = (-1)^{p+l} \mid f_b m_{f_b}, f_a m_{f_a}, lm_l \rangle. \tag{4.11}$$

In our experiment we are working with bosonic $^{87}$Rb atoms and we have $p = 0$. Furthermore, we always start the Feshbach association from two identical bosons, that is $f_a = f_b = 1$ and $m_{f_a} = m_{f_b} = 1$. From Eqs. (4.10) and (4.11) we can only create molecules with even rotation $l = 0, 2, 4, ....$ Indeed, in our experiment we produce Feshbach molecules with $l = 0$ [2]. The statements above imply that we have to change the initial state of one atom to produce Feshbach molecules with odd $l$. This is indeed the case and an example can be found in [Marte 2002]. (For a coupling between states with different $l$ and $m_l$ see [Dürr 2005].)

**Coupled basis.** In the coupled basis $\mid (f_a, f_b)f, m_f, lm_l \rangle$ we first have to calculate

$$(-1)^p P_{12} \mid (f_a, f_b)fm_f, lm_l \rangle =$$
$$(-1)^p P_{12} \sum_{m_{f_a}, m_{f_b}} \langle f_a m_{f_a}, f_b m_{f_b} \mid f m_f \rangle \mid f_a m_{f_a}, f_b m_{f_b}, lm_l \rangle. \tag{4.12}$$

---
[2] The closed channel connects to a state with $l_c = 0$. With the definition from [Chin 2009] we use an s-wave Feshbach Resonance.

## 4. Theory of diatomic molecules

As before, the parity operator causes a factor $(-1)^l$ in the ket involving $l$. To deal with the Clebsch-Gordon factor we use the following important identity [Edmonds 1960] (Eq. 4.13 will also become important when we consider the influence of nuclear statistics.)

$$\langle f_a m_{f_a}, f_b m_{f_b} \mid f m_f \rangle = (-1)^{f_a + f_b - f} \langle f_b m_{f_b}, f_a m_{f_a} \mid f m_f \rangle. \tag{4.13}$$

Afterwards, the uncoupling in the $(-1)^p P_{12}$ term is reversed and we end up with

$$\begin{aligned}(1 + (-1)^p P_{12}) \mid (f_a, f_b) f m_f, l m_l \rangle = \\ \mid (f_a, f_b) f m_f, l m_l \rangle + (-1)^{p+l+(f_a+f_b-f)} \mid (f_b, f_a) f m_f, l m_l \rangle.\end{aligned} \tag{4.14}$$

In analogy with the atomic basis we now obtain the rule $f + l = 0, 2, 4, \ldots$ if we start with identical Bosons in the states $f_a = f_b = 1$.

**Partially coupled molecular basis.** In analogy, we may consider the molecular basis of the form $\mid (s_a, s_b) S m_S, (i_a, i_b) I m_I, l m_l \rangle$. Using two Clebsch-Gordon factors to uncouple $S$ and $I$ we find:

$$\begin{aligned}(1 + (-1)^p P_{12}) \mid (s_a, s_b) S m_S, (i_a, i_b) I m_I, l m_l \rangle = \\ \mid (s_a, s_b) S m_S, (i_a, i_b) I m_I, l m_l \rangle + \\ (-1)^{p+l+(s_a+s_b-S)+(i_a+i_b-I)} \mid (s_b, s_a) S m_S, (i_b, i_a) I m_I, l m_l \rangle.\end{aligned} \tag{4.15}$$

In the case of identical $^{87}$Rb atoms with $s_{a,b} = 1/2$ and $i_{a,b} = 3/2$ this leads to $S+I+l = 0, 2, 4, \ldots$ This can be seen in Table 4.3, where I denote the kets in the atomic (molecular) basis with a subscript $\mid \ldots \rangle_a$ ($\mid \ldots \rangle_m$), respectively. The superscript $^s$ ($^a$) labels the (anti-)symmetrized states. As we deal with s-waves (no orbital angular momentum), the blocks show that the symmetric states only correspond to molecular states with $I + S$ even. These calculations imply that our Feshbach molecules do not rotate, that is $l = 0$. Table 4.3 is useful to transform wave-functions given in the atomic basis to the molecular basis. If one considers the table as a matrix, one can easily prove that it is unitary as expected.

**Fully-coupled molecular basis.** Finally, we also consider the case of the fully-coupled basis $\mid (SI) f l F M_F \rangle$ where **f** and **l** couple to form a total angular momentum **F**. Compared to the partially coupled molecular basis we need additional Clebsch-Gordon factors to show that $F = f + l = 0, 2, 4, \ldots$ for our $^{87}$Rb$_2$ Feshbach molecules.

**Exchange symmetries and nuclear spin.** Applying the Clebsch-Gordon argument from Eq. (4.13) we obtain the following restrictions on the nuclear spin function $\beta$ with respect to exchange of the nuclei (See Eq. (4.13); we replace each $f_{a,b}$ by $i$, and $f$ by $I$.): For antisymmetric wave-functions the prefactor has to be $-1 = (-1)^{2i-I}$. This depends on wether the nuclei obey Bose-Einstein or Fermi-Dirac statistic. For fermions, $i$ is half-integral and thus $2i$ is odd. Therefore, the total nuclear spin I has to be even i.e. $I = 2i - 1, 2i - 3, \ldots$ The other case can be derived similarly. This can be seen in Table 4.4.

Our rubidium atoms have nuclear spin $i = \frac{3}{2}$ and antisymmetric functions $\beta$ correspond to total nuclear spin $I = 0, 2$ while symmetric functions $\beta$ correspond to total nuclear spin $I = 1, 3$. We will use this result later on.

| $\lvert f_1, m_{f1}; f_2, m_{f2}\rangle_a$ $\lvert I, m_I, S, m_S\rangle_m$ | $\lvert a;a\rangle_a$ | $\lvert b;h\rangle_a^s$ | $\lvert a;g\rangle_a^s$ | $\lvert g;g\rangle_a$ | $\lvert f;h\rangle_a^s$ | $\lvert b;h\rangle_a^a$ | $\lvert a;g\rangle_a^a$ | $\lvert f;h\rangle_a^a$ |
|---|---|---|---|---|---|---|---|---|
| $\lvert 2,2,0,0\rangle_m$ | $\frac{\sqrt{3}}{4}$ | $\frac{1}{2}$ | $-\frac{1}{\sqrt{8}}$ | $-\frac{\sqrt{3}}{4}$ | $\frac{1}{2}$ | 0 | 0 | 0 |
| $\lvert 3,3,1,-1\rangle_m$ | $\frac{3}{4}$ | 0 | $\sqrt{\frac{3}{8}}$ | $\frac{1}{4}$ | 0 | 0 | 0 | 0 |
| $\lvert 3,2,1,0\rangle_m$ | $-\frac{\sqrt{3}}{4}$ | $\frac{1}{2}$ | $\frac{1}{\sqrt{8}}$ | $\frac{\sqrt{3}}{4}$ | $\frac{1}{2}$ | 0 | 0 | 0 |
| $\lvert 1,1,1,1\rangle_m$ | $-\frac{1}{\sqrt{40}}$ | $-\sqrt{\frac{3}{10}}$ | $\sqrt{\frac{3}{20}}$ | $-\frac{3}{\sqrt{40}}$ | $\sqrt{\frac{3}{10}}$ | 0 | 0 | 0 |
| $\lvert 3,1,1,1\rangle_m$ | $\sqrt{\frac{3}{80}}$ | $-\frac{1}{\sqrt{5}}$ | $-\sqrt{\frac{9}{40}}$ | $\sqrt{\frac{27}{80}}$ | $\frac{1}{\sqrt{5}}$ | 0 | 0 | 0 |
| $\lvert 2,2,1,0\rangle_m$ | 0 | 0 | 0 | 0 | 0 | $-\frac{1}{2}$ | $\frac{1}{\sqrt{2}}$ | $-\frac{1}{2}$ |
| $\lvert 3,2,0,0\rangle_m$ | 0 | 0 | 0 | 0 | 0 | $-\frac{1}{2}$ | $-\frac{1}{\sqrt{2}}$ | $-\frac{1}{2}$ |
| $\lvert 2,1,1,1\rangle_m$ | 0 | 0 | 0 | 0 | 0 | $\frac{1}{\sqrt{2}}$ | 0 | $-\frac{1}{\sqrt{2}}$ |

Table 4.3: Clebsch-Gordon coefficients. $\lvert ..\rangle_a$ denotes atomic basis whereas $\lvert ..\rangle_m$ denotes the molecular basis. The states $a$ to $h$ label the different hyperfine states with increasing energy. The symbols $\lvert ..\rangle^{s/a}$ denote the symmetrized and antisymmetrized wave-functions respectively.

### 4.6.2 Molecular symmetries

Symmetries not only reduce the number of Feshbach states as described in section 4.6.1, they also reduce the number of states in a Born-Oppenheimer potential. The latter will be explained in section 4.7. If we find a symmetry operator **O** which commutes with **H**, we can construct eigenfunctions which diagonalize both **H** and **O**. We can then label the eigenfunctions with the corresponding additional quantum numbers. In the case of a rotation this procedure leads to the orbital angular momentum quantum number and its projection used in section 4.2.

The following list includes those symmetry operators which we will use frequently later on.

1. **i** (parity): The first example is the operation which corresponds to an inversion of all electron and nuclear coordinates. States are called negative ("$P = -$") or positive ("$P = +$") depending on whether the wave-function changes sign or remains unaltered, respectively. The corresponding quantum number $P$ is exact in the sense defined on page 28.

2. $C_2(\boldsymbol{n})$ (rotation through $\pi$): Rotation of all electrons and nuclei about the $\boldsymbol{n}$-axis through an angle $\pi$. If we take the $\boldsymbol{n}$-axis perpendicular to the internuclear axis (For example the $y$-axis if we take the $z$-axis as internuclear axis.) this operation has no influence on the electronic wave-function as only relative positions with respect to the nuclei enter in the electronic wave-function.

3. $\sigma_{el}(x,z)$ (reflection; As the inversion **i** it is also called parity. See [Lefebvre-Brion/Field 2004] page 139): Reflection of electron and nuclear coor-

## 4. Theory of diatomic molecules

| Co-ordinate function $\Psi$ | Nuclear spin function $\beta$ | Statistical weight | Nuclear spin I | Total function $\Psi^T$ | For Bose statistics of the nuclei | For Fermi statistics of the nuclei |
|---|---|---|---|---|---|---|
| sym. | sym. | $(2i+1)(i+1)$ | $2i, 2i-2, ...$ | sym. | occurs | – |
| sym. | antisym. | $(2i+1)i$ | $2i-1, 2i-3, ...$ | antisym. | – | occurs |
| antisym. | sym. | $(2i+1)(i+1)$ | $2i, 2i-2, ...$ | antisym. | – | occurs |
| antisym. | antisym. | $(2i+1)i$ | $2i-1, 2i-3, ...$ | sym. | occurs | – |

Table 4.4: Symmetry with respect to exchange of nuclei. The table shows the symmetry of the eigenfunctions including nuclear spin for homonuclear diatomic molecules. One should keep in mind that $2i$ can be an odd number if $i$ is half integral. For the definitions of the wave-functions see Eq. 4.7. Adapted from [Herzberg 1950].

dinates with respect to a plane, containing the internuclear axis. As we have $\mathbf{i} = \boldsymbol{C}_2(y) \cdot \boldsymbol{\sigma}_{el}(x,z)$ only the $\sigma_{el}$ operation determines the state. The states are then denoted $\Sigma^+$ or $\Sigma^-$ according to the positive or negative result of the $\sigma_{el}$ operation, respectively (See also 2. above and [Hougen 1970, Nikitin 1994])

4. $\mathbf{i_e}$ (electronic inversion): This symmetry operation only leads to good quantum numbers for homonuclear diatomic molecules. It corresponds to an inversion of the electronic coordinates. States are called **u**ngerade or **g**erade depending on whether the wave-function changes sign (eigenvalue $\iota_e = -1$) or does not change sign (eigenvalue $\iota_e = +1$).

5. $\mathbf{i_n}$ (nuclear exchange): Inversion of nuclear coordinates. This symmetry operation only leads to good quantum numbers for homonuclear molecules. The electronic states are called **s**ymmetric and **a**symmetric corresponding to the different eigenvalues $\iota_n = +1$ and $\iota_n = -1$.

All operations are given with respect to molecule fixed coordinates. For details on the implementation of inversion with Euler angles $\phi$ and $\theta$ in a diatomic molecule see [Kronig 1930, Hougen 1970]. When looking back at the Born-Oppenheimer product function in Eq. (4.7) it should be clear that these symmetries only operate on the wave-function $\Psi$ (the label $n$ is not important at the moment and will be suppressed in the following.), which includes the electron spin (compare Table 4.4). The symmetry of the total wave-function $\Psi^T$ is determined as soon as the statistics of the nucleus is known. An overview of the different exchange symmetries is given in Table 4.4. The so-called coordinate function $\Psi$ includes everything but nuclear spin. Its symmetry can be inferred from the electronic term symbols, as I will discuss in the next section for the example of the ground state and the first excited state. As before, the nuclear spin function is denoted with $\beta$ where the degeneracy corresponding to the $m_I$ quantum number is given in the third column.

# 4. Theory of diatomic molecules

## 4.7 Symmetries and Hund's cases in selected $^{87}$Rb$_2$ states

When we discuss the experimental results, only few states are relevant. These states are discussed in detail below. The argumntation could then be applied to different coupling schemes or states with different symmetries[3].

### 4.7.1 The electronic ground state connecting to the 5S+5S asymptote

The diatomic ground state is a $5^2S_{1/2} + 5^2S_{1/2}$ state. The total electron spin $S$ can be either 1 (triplet state) or 0 (singlet state). Let us first consider molecules in the $a\,^3\Sigma_u^+$ triplet state. When we employ the operator identity $\mathbf{i} = \boldsymbol{\sigma}_{el}(x,z) \cdot \boldsymbol{C}_2(y)$ we can show that the rotational ground state has parity "+" and electronic inversion quantum number "u":

$$\mathbf{i} \cdot \Psi^+ = (\boldsymbol{\sigma}_{el}(x,z) \cdot \boldsymbol{C}_2(y)) \cdot \Psi^+$$
$$= \boldsymbol{\sigma}_{el}(x,z) \cdot \Psi^+ = +\Psi^+ \quad (4.16)$$
$$\mathbf{i_e} \cdot \Psi_u = -\Psi_u.$$

Here and in the following discussion variables or symmetry labels are not specified if they are not used. Furthermore, our rotational ground-state in the $a\,^3\Sigma_u^+$ potential is asymmetric ("a") with respect to nuclear exchange. This can be seen as follows from the operator identity $\mathbf{i_n} = \mathbf{i_e} \cdot \mathbf{i}$

$$\mathbf{i_n} \cdot \Psi_u^+ = (\mathbf{i_e} \cdot \mathbf{i}) \cdot \Psi_u^+ = -\Psi_u^+. \quad (4.17)$$

Now we make use of the fact that we are dealing with $\Sigma$ states which obey a Hund's coupling case b) scheme. In this case, the spin function $\chi$ is "symmetric" that means it does not change sign under the operations $\boldsymbol{\sigma}_{el}(x,z)$ and $\boldsymbol{C}_2(y)$. This is a consequence of the spin functions's independence on the orientation of the molecule (See section 4.5). A rigorous proof is not trivial and can be found in [Kronig 1930]. If we also include the results of Table 4.4 we end up with the following restrictions for the rotational ground state[4]:

| $\Psi$ | $\beta$ | I | $\Psi^T$ |
|---|---|---|---|
| s | antisymm. | 0,2 | antisymm |
| a | symm. | 1,3 | antisymm |

Here, we have made use of the fact that our nuclei obey Fermi statistics ($i=3/2$). Thus, the total molecular wave-function $\Psi^T$ has to be antisymmetric with respect to nuclear exchange.

All together we arrive at Table 4.5 which shows the rotational quantum numbers and their symmetries for the low-lying rotational levels. From this table it is obvious that

---
[3][Herzberg 1950] gives several examples including different electronic states.
[4]The symmetry of the rotational wave-function $\Psi^{rot}$ is $(-1)^N$. Even in the presence of interaction where the Born-Oppenheimer product ansatz fails, this statement remains true. See footnote on page 138 of [Herzberg 1950].

45

4. Theory of diatomic molecules

| $a\,^3\Sigma_u^+$ | $N$ | 0 | 1 | 2 | 3 | 4 | 5 |
|---|---|---|---|---|---|---|---|
| | $P$ | + | - | + | - | + | - |
| | $\iota_n$ | a | s | a | s | a | s |
| | $I$ | 1,3 | 0,2 | 1,3 | 0,2 | 1,3 | 0,2 |
| $X\,^1\Sigma_g^+$ | $N$ | 0 | 1 | 2 | 3 | 4 | 5 |
| | $P$ | + | - | + | - | + | - |
| | $\iota_n$ | s | a | s | a | s | a |
| | $I$ | 0,2 | 1,3 | 0,2 | 1,3 | 0,2 | 1,3 |

Table 4.5: Symmetries of the low-lying rotational and vibrational levels in the electronic ground-state connecting to the $5S_{1/2}+5S_{1/2}$ asymptote. As above $P$ denotes the parity, $\iota_n$ is the quantum number for nuclear exchange and $I$ is the total nuclear spin.

these symmetries are also consistent for states at the dissociation threshold, for example our Feshbach state. This is important because the Feshbach state is a mixture of singlet and triplet states.

### 4.7.2 The first electronic excited $(1)\,^3\Sigma_g^+$ state

The excited state $|e\rangle$ which we studied in detail using one-photon spectroscopy and which we employ for our Raman process is located in the $(1)\,^3\Sigma_g^+$ $(5^2S_{1/2} + 5^2P_{1/2})$ potential. In contrast to the previously discussed $a\,^3\Sigma_u^+$ ground state, we now find a strong second-order spin-orbit and direct spin-spin interactions. As already mentioned, these interactions cause a strong coupling of $\mathbf{S}$ to the internuclear axis (For a derivation see section 5.3.1.) and force the states to show strong Hund's coupling case c) features. The spin-spin interaction causes a huge splitting between the $1_g$ and $0_g^-$ states, where we use the nomenclature of Table 4.2. Now, we explore the symmetries in the excited state. As we already know how to proceed in the Hund's case b) coupling scheme, we study the transition from case b) to case c). The correlations between the rotational levels in the two rotational ladders are shown in Fig. 4.5.

The left side of Fig. 4.5 shows the rotational ladder for triplet molecules as discussed in section 4.5. The rotational energies are proportional to $N(N+1)$ and the parity varies as $(-1)^N$ because we have a $\Sigma^+$ state. The latter means that the electronic wave-function does not change sign under $i$ (Compare section 4.6.2). The state with $J = 0$ on the left side shows quantum numbers $N = 1$ and "-" parity. This state has to correlate with the rotational ground state in the Hund's case c) scheme and gives rise to the name $0_g^-$. As explained above, the "g" label denotes the quantum number for the exchange of electrons. It only exists for homonuclear molecules and is not affected by the order of coupling. For the $0_g^-$ state with $J = 1$ we find three candidates. However, one of these states has $N = 2$ and is ruled out due to its "+" parity (See page 43.). One of the two remaining states has the "wrong" energy. We conclude that the symmetries

4. Theory of diatomic molecules

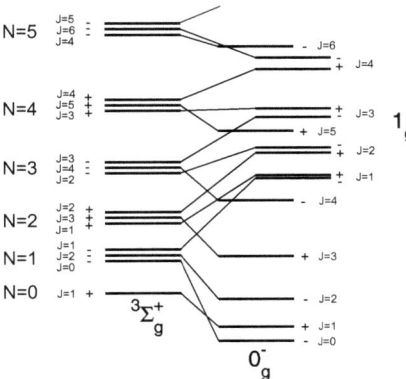

Figure 4.5: Transition from Hund's case b) (left) to c) for (1) $^3\Sigma_g^+$ states (adapted from [Herzberg 1950]). The scheme shows the rotational structure in both cases and how they connect. In both cases, $J$ corresponds to the total angular momentum, $\pm$ is the parity (See section 4.6.2), and $N$ is the rotation of the molecule excluding the total electron spin $S$.

of the rotational ground state fix the rotational ladder in the $0_g^-$ case. It is clear that the remaining states have to connect to the $1_g$ state. Ordering the remaining states with increasing $J$ we observe that each state exists twice, once with negative and once with positive parity. These two levels are degenerate if the spin-spin and spin-rotation interactions are small. For high $J$, the splitting between the two degenerate states increases because the interaction between total angular momentum $J$ and the total orbital angular momentum $L$ becomes stronger. In the literature this is known as $\lambda$-type doubling (See for example [Lefebvre-Brion/Field 2004] page 214.).

We are now prepared to set up the table for the $0_g^-$ and $1_g$ states separately. From Fig. 4.5 we know that the rotational ground state has negative parity. As in formula (4.16) we now obtain

$$\mathbf{i} \cdot \Psi^- = (\boldsymbol{\sigma}_{el}(x,z) \cdot \boldsymbol{C}_2(y)) \cdot \Psi^-$$
$$= \boldsymbol{\sigma}_{el}(x,z) \cdot \Psi^- = -\Psi^- \quad (4.18)$$
$$i_e \cdot \Psi_g = +\Psi_g.$$

which leads immediately to an asymmetric ("a") eigenvalue with respect to nuclear exchange because

$$\mathbf{i}_n \cdot \Psi_g^- = (\mathbf{i_e} \cdot \mathbf{i}) \cdot \Psi_g^- = -\Psi_g^-. \quad (4.19)$$

Thus we get the third column in Table 4.6. As the nuclear spin is only weakly coupled to the rotation, the parity and therefore the a/s and $I$ quantum numbers change as $(-1)^J$.

## 4. Theory of diatomic molecules

| | J | 0 | 1 | 2 | 3 | 4 | 5 |
|---|---|---|---|---|---|---|---|
| $(1)\,^3\Sigma_g^+$ | P | − | + | − | + | − | + |
| $0_g^-$ | $\iota_n$ | a | s | a | s | a | s |
| | I | 1,3 | 0,2 | 1,3 | 0,2 | 1,3 | 0,2 |
| | J | | 1 | 2 | 3 | 4 | 5 |
| $(1)\,^3\Sigma_g^+$ | P | | +   − | +   − | +   − | +   − | +   − |
| $1_g$ | $\iota_n$ | | s   a | s   a | s   a | s   a | s   a |
| | I | | 0,2  1,3 | 0,2  1,3 | 0,2  1,3 | 0,2  1,3 | 0,2  1,3 |

Table 4.6: Symmetries of the deeply bound, low-lying rotational levels in the first excited $(1)\,^3\Sigma_g^+$ state. $P$ denotes the parity, $\iota_n$ is the quantum number for nuclear exchange and $I$ is the total nuclear spin. In contrast to the ground state, the rotation energy is now proportional to $J(J+1)$. Furthermore, we have to treat the $0_g^-$ and $1_g$ states separately as they have different parities.

Here, the crucial point is that the electronic spin is coupled to the internuclear axis and contributes to the total angular momentum. As already mentioned, this destroys the $N$ quantum number [Herzberg 1950]. For the second row I only have to mention that each $J$-level is twofold degenerate and that $J \geq \Sigma = 1$. The rest can be calculated as for the $0_g^-$ state. In principle, the rotational ladder does not end with $J = 6$, although we do not observe levels with values of $J \geq 4$ in our experiment. This will be discussed in detail in chapter 6.

### 4.8 Selection rules for bound-bound transitions

Up to now we have only considered bare molecular states without any light field. The symmetries then restricted the number of molecular levels. If we illuminate the molecules with coherent light, we are not able to couple arbitrary molecular states. This is due to the dipole matrix element connecting the initial and the final states which has to be nonzero for an optical dipole transition. These selection rules will be the topic of the present section. To get an overview of the involved quantities in one- and two-photon spectroscopy, we first consider the general transition rate $W_{fi}$ of an electric one-photon process

$$W_{fi} \propto |\langle f | \boldsymbol{\epsilon} \cdot \boldsymbol{\mu} | i \rangle|^2. \tag{4.20}$$

Here, we use the initial state $|i\rangle$ and and the final state $|f\rangle$ and $\boldsymbol{\epsilon}$ is a unit vector specifying the polarization of the electromagnetic field. The operator $\boldsymbol{\mu}$ is the sum of all electronic dipoles, that is

$$\boldsymbol{\mu} = e \sum_{k=1}^{n} \mathbf{r}_k, \tag{4.21}$$

# 4. Theory of diatomic molecules

where we use the position of the $k$th electron $\mathbf{r}_k$. Thus, $\boldsymbol{\epsilon}\cdot\boldsymbol{\mu}$ is the electric dipole operator and it is clear how we can in principle treat the one-photon transition.

In general, there are two types of selection rules which follow from Eq. (4.20). The first class applies to all transitions no matter which Hund's coupling case the initial and final state belong to. The second type depends on the coupling case and one has to treat each combination separately. The general selection rules for $\boldsymbol{\mu}$ are:

1. Symmetry selection rules.

$$+ \leftrightarrow - \qquad \text{(parity; } \boldsymbol{\sigma}_{el}(x,z) \text{ in the laboratory fixed system),} \qquad (4.22)$$
$$g \leftrightarrow u \qquad \text{(inversion; } i \text{ in the molecule fixed system),} \qquad (4.23)$$
$$s \leftrightarrow s, \quad a \leftrightarrow a \qquad \text{(exchange of nuclei; } C_2 \text{ in the laboratory fixed system).} \qquad (4.24)$$

These selection rules hold in any Hund's case because they involve exact quantum numbers. In contrast,

$$\Sigma^\pm \leftrightarrow \Sigma^\pm \qquad \text{(parity; } \boldsymbol{\sigma}_{el}(x,z) \text{ in the molecule fixed system)}$$

only holds if $\Sigma$ is a good quantum number, that is in Hund's case a). These selection rules are treated in [Herzberg 1950] on page 240 and follow from the independence of the dipole matrix element under the symmetry operations[5]. Interestingly, the selection rule for a/s-states also holds in collisional processes.

2. Selection rules for vibrational wave-functions. Neglecting the $R$-dependence of all Born-Oppenheimer-product functions except $\Psi^{vib}$, one factor of the matrix elements in Eq. (4.20) is the so-called Franck-Condon factor

$$\left|\langle \Psi_f^{vib} | \Psi_i^{vib} \rangle\right|^2, \qquad (4.25)$$

where we use the vibrational wave-functions $\Psi_{f,i}^{vib}$ for the initial and final state as defined in Eq. (4.4).

The last selection rule involves the rotational part of the Born-Oppenheimer-product function.

3. Angular momentum selection rules. This matrix element will be discussed in detail in the next section. I will give the results for the reader who is not interested in the details.

$$\begin{aligned} \Delta\Lambda &= 0, \pm 1, \\ \Delta S &= 0, \\ \Delta I &= 0, \\ \Delta m_F &= 0, \pm 1, \\ \Delta F &= 0, \pm 1. \end{aligned} \qquad (4.26)$$

---

[5] A detailed description can be also found in the books of [Hougen 1970], page 28-30 and [Lefebvre-Brion/Field 2004], page 139.

## 4. Theory of diatomic molecules

$\Delta\Lambda = 0, \pm 1$ follows from angular momentum conservation and $\Delta S = 0$ holds because the dipole operator does not involve any spins. The total nuclear spin $I$ obeys $\Delta I = 0$. Due to the laser polarizations in our experiments only $\pi$ transitions are allowed and thus $\Delta m_F = 0$. The last selection rule contains the total angular momentum and reads $\Delta F = 0, \pm 1$. If a quantum number, for example $F$, becomes bad, the states are a superposition of different $|F\rangle$ states and levels with higher $F$ can be observed. This is crucial for explaining the spectra in the excited state.

Up to now I only discussed one-photon transitions and one could wonder if there is anything special about two-photon (Raman) transitions. In principle no, we only have another transition rate. In the simplest case where we neglect the laser noise and decoherence processes, we just plug the transition moments into a three-level model as described by Winkler [Winkler 2007a]. If one would like to include the noise and other decoherence processes appropriately, one has to use Lindblad-type equations for the corresponding density matrices as described in [Walls/Milburn 1994, Lang 2008].

### 4.8.1 Derivation of the selection rules including nuclear spin

In this section I will give a derivation of the selection rules listed above [Lysebo 2010] because the evaluation of the dipole matrix elements is a little more complicated when the nuclear spin is included. The derivation basically consists of five steps:

(i) Set up dipole matrix element using tensor operators.

(ii) Transform the dipole operator to the molecule-fixed coordinate system.

(iii) Separate the Franck-Condon factors.

(iv) Choose appropriate basis for the evaluation of the rotational part of the remaining operator and then transform the initial and final state to this basis. In this step one has also to separate the rotational wave-function.

(v) Reconnect everything and use the fact that we use $\pi$-light.

As we are interested in the dipole matrix element $\langle f\,|\boldsymbol{\epsilon}\cdot\boldsymbol{\mu}|\,i\rangle$, we have to consider the dipole operator $\boldsymbol{\epsilon}\cdot\boldsymbol{\mu}$. In Cartesian coordinates, $\boldsymbol{\epsilon}$ is a unit vector specifying the polarization of the electromagnetic field and $\boldsymbol{\mu}$ is given in Eq. 4.21. Again, $|\,i\rangle$ and $|\,f\rangle$ are the initial and final state, respectively. This calculation (as the rest of the calculations) is performed most easily using spherical tensor operators [Edmonds 1960, Rose 1957, Brink/Satchler 1971] and [Lindner 1984].

**Step (i)** In this calculus the $z$-component corresponds to the 0th tensor component, that is $T_0^{(1)}(\boldsymbol{\epsilon})$ and the $x$ and $y$ components are given as a linear combination of the two operators $T_1^{(1)}(\boldsymbol{\epsilon})$ and $T_{-1}^{(1)}(\boldsymbol{\epsilon})$. Using the formula for scalar products in the realm of tensor operators we obtain

$$d_{fi} = \langle f \mid \boldsymbol{\epsilon} \cdot \boldsymbol{\mu} \mid i \rangle = \sum_p (-1)^p \cdot \langle f \mid T^{(1)}_{-p}(\boldsymbol{\epsilon}) \cdot T^{(1)}_p(\boldsymbol{\mu}) \mid i \rangle.$$

**Step (ii)** The next step is to transform the space-fixed component of the dipole operator $\boldsymbol{\mu}$ into a molecule-fixed system. This step is necessary to cope with the electronic wavefunctions which are given in a molecule-fixed system.

$$T^{(1)}_p(\boldsymbol{\mu}) = \sum_m D^{(1)}_{p,\,m}(\omega)^* \cdot T^{(1)}_m(\boldsymbol{\mu}),$$

where the index $m$ always denotes molecule-fixed components. If we now use the fact that $T^{(1)}_{-p}(\boldsymbol{\epsilon})$ is a constant scalar and put the two equations together we arrive at

$$\langle f \mid \boldsymbol{\epsilon} \cdot \boldsymbol{\mu} \mid i \rangle = \sum_{p,\,m} (-1)^m \cdot T^{(1)}_{-p}(\boldsymbol{\epsilon}) \cdot \langle f \mid D^{(1)}_{p,\,m}(\omega)^* \cdot T^{(1)}_m(\boldsymbol{\mu}) \mid i \rangle.$$

Up to now, the scheme is general and applies to each one-photon transition in a molecule. Further evaluation depends on the initial and final states, as we now have to decompose them into a rotational part (upon which the D-matrix acts) and a spin part. This means that we have to choose a common basis. If the choice was appropriate for the problem, at least one state is close to a basis vector and we don't have to evaluate several sums. In the singlet and triplet ground states, the rotational energy ladder is proportional to $N(N+1)$ and a Hund's coupling scheme b) is appropriate. In contrast the excited state is closer to Hund's case c). With the decomposition of $\mid i \rangle$ into electronic, vibrational, and rotational wave-function we obtain

$$\mid i \rangle = \mid \Psi^{vib}_i \rangle \mid \Psi^e_i \cdot \Psi^{rot}_i \rangle.$$

The same holds for $\mid f \rangle$. The matrix elements can then be written

$$\begin{aligned}d_{fi} = \langle f \mid \boldsymbol{\epsilon} \cdot \boldsymbol{\mu} \mid i \rangle = \sum_{p,\,m} (-1)^m \cdot T^{(1)}_{-p}(\boldsymbol{\epsilon}) \cdot \langle \Psi^{vib}_f \mid \Psi^{vib}_i \rangle \cdot \\ \cdot \langle \Psi^e_f \cdot \Psi^{rot}_f \mid D^{(1)}_{p,\,m}(\omega)^* \cdot T^{(1)}_m(\boldsymbol{\mu}) \mid \Psi^e_i \cdot \Psi^{rot}_i \rangle.\end{aligned} \quad (4.27)$$

In this step we used Eq. (4.21) which implies that the matrix elements are independent of $\Psi^{vib}$. In the following discussion we ignore the Franck-Condon factor $\langle \Psi^{vib}_f \mid \Psi^{vib}_i \rangle$ as it is the same within one vibrational level.

Our initial state can be described as a superposition of different molecular states, (See quantum numbers for the Feshbach state and Table 4.3.) and we represent it as a superposition of molecular states

$$\mid i \rangle = \sum_r c_r \mid qSm_S Im_I lm_l \rangle_r, \quad (4.28)$$

## 4. Theory of diatomic molecules

which includes the electronic wave-function $|\Psi_i^e\rangle$ and the rotational wave-function $|\Psi_i^{rot}\rangle$. The subscript $r$ denotes the fraction of the different states. Our final excited state is deeply bound and can be written in the form

$$|f\rangle = \sum_s c'_s \, |\, q'S'\Omega'I'\Omega_{I'}F'M'\Omega_{F'}\rangle_s. \qquad (4.29)$$

In the following, we will omit the sum over $r$ and $s$ and evaluate each matrix element separately. Furthermore, we omit the label $q$, $q'$ for extra quantum numbers which specifies the electronic state and the vibrational quantum number.

**Transformation for the final state.** With the abbreviation $\omega = (\pi/2, \theta, \phi)$ the rotation matrix becomes $D^{(F)}_{M_F\Omega_F}(\omega)$. It acts as an eigenfunction for the absolute value of the total angular momentum $\mathbf{F}^2$, its projection on the space-fixed axis $F_z$, and its molecule-fixed projection $F_Z$ [Edmonds 1960]. We use a product ansatz for the molecular wave-function to get

$$\langle \omega \,|\, S'\Omega'I'\Omega'_I F'M'_F\Omega'_F\rangle = \sqrt{\frac{2F'+1}{4\pi}} D^{(F')}_{M'_F\Omega'_F}(\omega)^* \cdot |\, S'\Omega'I'\Omega'_I\rangle.$$

**Transformation for the initial state.** To further develop the expression, we need to consider the molecular Hund's case a) states because the electronic wave-functions are given in the molecule-fixed system. The electronic state $|\,Sm_S\rangle$ expanded in terms of the molecule-fixed spin states $|\,S\Sigma\rangle$ is

$$\langle \omega \,|\, SM_S\rangle = \sum_\Sigma D^{(S)}_{M_S\Sigma}(\omega)^* \,|\, S\Sigma\rangle,$$

and for $|\, Im_I\rangle$:

$$\langle \omega \,|\, IM_I\rangle = \sum_{\Omega_I} D^{(I)}_{M_I\Omega_I}(\omega)^* \,|\, I\Omega_I\rangle.$$

Finally, for states which connect to the two-atom asymptote we obtain

$$\langle \omega \,|\, lm_l\rangle = \sqrt{\frac{2l+1}{4\pi}} D^{(l)}_{m_l 0}(\omega)^*.$$

Putting all pieces together we can write:

$$\langle \omega \,|\, SM_S IM_I lm_l\rangle = \sum_{\Sigma\Omega_I} \sqrt{\frac{2l+1}{4\pi}} D^{(l)}_{m_l 0}(\omega)^* D^{(I)}_{M_I\Omega_I}(\omega)^*$$
$$\cdot D^{(S)}_{M_S\Sigma}(\omega)^* \,|\, S\Sigma I\Omega_I\rangle.$$

The product of two different rotation matrices is given by:

$$D^{(l)}_{m_l 0}(\omega)^* D^{(S)}_{M_S\Sigma}(\omega) = \sum_J (2J+1) \cdot$$
$$\cdot \begin{pmatrix} S & l & J \\ M_S & m_l & -M_J \end{pmatrix} \begin{pmatrix} S & l & J \\ \Sigma & 0 & -\Omega \end{pmatrix} D^{(J)}_{-M_J -\Omega}(\omega)^*.$$

Now, $D^{(J)}_{-M_J-\Omega}(\omega)^* = (-1)^{\Omega-M_J} D^{(J)}_{M_J\Omega}(\omega)^*$ [Edmonds 1960]. Similar considerations for the product $D^{(J)}_{M_J\Omega}(\omega)^* D^{(I)}_{M_I\Omega_I}(\omega)^*$ lead in total to

$$\langle \omega \mid SM_S IM_I lm_l \rangle = \sum_\Sigma \sum_{\Omega_I} \sum_{J,F} \sqrt{\frac{2l+1}{4\pi}}(2J+1)(2F+1)(-1)^{\Omega-M_J}(-1)^{\Omega_F-M_F}$$
$$\begin{pmatrix} S & l & J \\ M_S & m_l & -M_J \end{pmatrix} \begin{pmatrix} S & l & J \\ \Sigma & 0 & -\Omega \end{pmatrix} \begin{pmatrix} I & L & F \\ M_I & m_J & -M_F \end{pmatrix}$$
$$\begin{pmatrix} I & J & F \\ \Omega_I & \Omega & -\Omega_F \end{pmatrix} D^{(F)}_{M_F-\Omega}(\omega)^* \mid S\Sigma I\Omega_I \rangle$$

(4.30)

**Final evaluation of $d_{fi}$.** Going back to Eq. (4.27) we can finally average over all Euler angles to obtain:

$$d_{fi} = \sum_{p,m} (-1)^m \cdot T^{(1)}_{-p}(\boldsymbol{\epsilon}) \cdot \sum_\Sigma \sum_{\Omega_I} \sum_{J,F} \sqrt{\frac{2l+1}{4\pi}}(2J+1)(2F+1)(-1)^{\Omega-M_J+\Omega_F-M_F}.$$
$$\begin{pmatrix} S & l & J \\ M_S & m_l & -M_J \end{pmatrix} \begin{pmatrix} S & l & J \\ \Sigma & 0 & -\Omega \end{pmatrix} \begin{pmatrix} I & J & F \\ M_I & m_J & -M_F \end{pmatrix}.$$
$$\begin{pmatrix} I & J & F \\ \Omega_I & \Omega & -\Omega_F \end{pmatrix} \sum_m \langle S'\Omega'I'\Omega'_I \mid T^{(1)}_m(\boldsymbol{\mu}) \mid S\Omega I\Omega_I \rangle \cdot$$
$$\int D^{(F)}_{M'_F-\Omega'}(\phi,\theta,0)^* D^{(F)}_{M_F-\Omega}(\phi,\theta,0) D^{(1)}_{p,m}(\phi,\theta,0)^* \sin\theta \, d\phi \, d\theta.$$

(4.31)

The integral gives

$$8\pi^2 (-1)^{\Omega_F-M_F} \begin{pmatrix} F' & 1 & F \\ -\Omega'_F & m & \Omega_F \end{pmatrix} \begin{pmatrix} F' & 1 & F \\ -M'_F & p & M_F \end{pmatrix}$$

(4.32)

and we arrive at

$$d_{fi} = \sum_{p,m} (-1)^m \cdot T^{(1)}_{-p}(\boldsymbol{\epsilon}) \cdot \sum_\Sigma \sum_{\Omega_I} \sum_{J,F} 8\pi^2 \sqrt{\frac{2l+1}{4\pi}}(2J+1)(2F+1)(-1)^{\Omega-M_J+\Omega_F-M_F}$$
$$(-1)^{\Omega'_F-M'_F} \begin{pmatrix} S & l & J \\ M_S & m_l & -M_J \end{pmatrix} \begin{pmatrix} S & l & J \\ \Sigma & 0 & -\Omega \end{pmatrix} \begin{pmatrix} I & J & F \\ M_I & m_J & -M_F \end{pmatrix}$$
$$\begin{pmatrix} I & J & F \\ \Omega_I & \Omega & -\Omega_F \end{pmatrix} \sum_m \langle S'\Omega'I'\Omega'_I \mid T^{(1)}_m(\boldsymbol{\mu}) \mid S\Omega I\Omega_I \rangle$$
$$\begin{pmatrix} F' & 1 & F \\ -\Omega'_F & m & \Omega_F \end{pmatrix} \begin{pmatrix} F' & 1 & F \\ -M'_F & p & M_F \end{pmatrix}$$

(4.33)

## 4. Theory of diatomic molecules

For $\pi$ light, $p = 0$ holds and one immediately obtains $M'_F = M_F$. Since the electric dipole operator $T_m^{(1)}(\boldsymbol{\mu})$ is independent of spin, we must have $S' = S$, $\Omega' = \Omega$, $I' = I$, and $\Omega'_I = \Omega_I$. However $\Omega'_F = \Omega'_I + \Omega'$ so that $\Omega'_F = \Omega_F$ and hence $m = 0$. Furthermore, $F' = F, F \pm 1$. Of course, some other matrix elements can be zero because of symmetry-rules for the Clebsch-Gordon coefficients.

# 5 One-photon spectroscopy of the $(1)\,^3\Sigma_g^+$ potential

Progress in the field of ultracold atomic and molecular gases has been strongly linked to developments in molecular spectroscopy. Photoassociation spectroscopy, for example, has been important for the studies of ultracold atomic collisions and production of ultracold molecules [Weiner 1999, Jones 2006, Köhler 2006, Sage 2005]. In 2008, after carrying out spectroscopic searches, several groups managed to produce cold and dense samples of deeply bound molecules in well defined quantum states [Lang 2008, Ni 2008, Danzl 2008, Viteau 2008, Deiglmayr 2008, Ospelkaus 2010]. For this, a variety of optical transfer and filtering schemes were developed which involved electronically excited molecular levels. These levels had to be properly chosen for high efficiency and selectivity of the production of molecules. Very recent work [Bai 2011] investigated the spin-orbit coupled $A\,^1\Sigma_u^+$ and $b\,^3\Pi_u$ states of Cs$_2$. In other work a detailed analysis of weakly bound Rb$_2$ levels of the excited $1_g$ state close to the $^5S_{1/2}+^5P_{1/2}$ dissociation limit is currently under way [Bergeman 2011] (see also related work in Fig. 13 of [Jones 2006]).

I present our measurements and analysis for deeply bound ($v = 0...15$) levels of the $(1)\,^3\Sigma_g^+(5S_{1/2} + 5P_{1/2})$ potential of $^{87}$Rb$_2$ [Takekoshi 2011]. The energy of this potential as a function of the internuclear distance $R$ is shown in Fig. 5.4 at page 62, together with other nearby potentials. These states are important, as they contribute binding energies via higher order perturbation theory. This includes second-order spin-orbit coupling which I explain in section 5.3. The states in the $(1)\,^3\Sigma_g^+$ potential are relevant for the production of deeply bound molecules in the $a\,^3\Sigma_u^+$ state via stimulated Raman adiabatic passage (STIRAP) [Lang 2008, Winkler 2007a]. The levels of the $a\,^3\Sigma_u^+$ potential have been mapped out and identified in detail in a recent publication [Strauss 2010]. (See also chapter 6.) The $(1)\,^3\Sigma_g^+$ potential is not easily accessible for spectroscopy in conventional setups since the molecules in ordinary Rb$_2$ gas samples are found in their singlet ground state $X\,^1\Sigma_g^+$. From the ground state the $(1)\,^3\Sigma_g^+$ potential cannot be reached with an optical transition due to the selection rule $\Delta S = 0$. In realm of symmetry this transition corresponds to a change from "g" to "u" symmetry which is forbidden in a two-photon process as I explained in section 4.8. Only recently, noteworthy experimental investigations of the Rb$_2$ $(1)\,^3\Sigma_g^+$ potential were carried out. Lozeille et al. [Lozeille 2006] performed photoionization spectroscopy with ultracold molecules starting from a magneto-optical trap to resolve the large $0_g^- - 1_g$ splitting of the vibrational levels. Mudrich et al. [Mudrich 2010] used pump-probe photoionization spectroscopy of Rb$_2$ formed on helium nanodroplets to measure the vibrational progression of deeply bound levels. Our work goes well beyond these measurements as we fully resolve the rotational, hyperfine, and Zeeman structure with an absolute accuracy as high as 60 MHz.

The starting point of our spectroscopy is an ultracold ensemble of weakly bound Rb$_2$

# 5. One-photon spectroscopy of the (1) $^3\Sigma_g^+$ potential

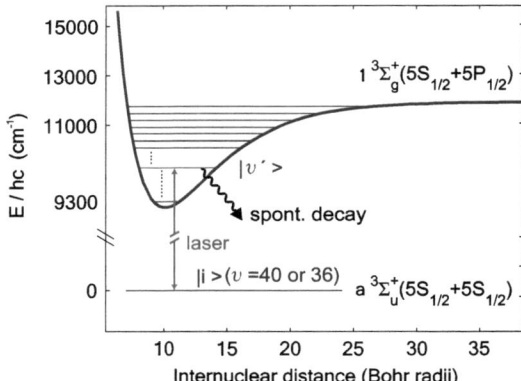

Figure 5.1: Molecular levels of $^{87}$Rb$_2$ in the (1) $^3\Sigma_g^+$ potential. A tunable laser couples the molecular level $|i\rangle$ and the excited level $|e\rangle$. The laser can be tuned to almost any deeply bound level of the (1) $^3\Sigma_g^+$ potential. Level positions are detected through resonantly enhanced loss of Feshbach molecules.

Feshbach molecules in a well defined quantum state. A tunable laser with sub-MHz linewidth drives a one-photon transition to individual levels in the (1) $^3\Sigma_g^+$ potential. We obtain loss spectra for various magnetic fields ranging from 0 G to about 1000 G. The data of one vibrational level is well explained with an effective Hamiltonian which contains terms for molecular rotation as well as spin-spin, hyperfine and Zeeman interactions.

This chapter is organized as follows: Section 5.1 presents the experimental setup including the method of "adiabatic transfer over avoided crossings". In section 5.2 I present the setup for the one-photon spectroscopy including the technique we use to go to lower magnetic fields. Afterwards, I give a short summary of the different terms we use in the Hamiltonian (Section 5.3). Sections 5.4.1 and 5.4.2 present the results and explain the level structure including Zeeman shifts in detail.

## 5.1 Experimental setup

Our one-photon spectroscopy at 986.8 G works as follows. The Feshbach molecules[1] in state $|i\rangle$ are irradiated by a laser ($\approx 100\mu$W) which corresponds to a Rabi frequency $\Omega$ on the order of $2\pi \times 0.1$ MHz (Fig. 5.1). The light pulses are rectangular and typically last for 50 ms. If we hit an $|i\rangle - |e\rangle$ resonance, our laser induces losses of Feshbach molecules due to the short lifetime of the excited state. After the laser light is switched off, we measure

---
[1]The creation of Feshbach molecules is described in detail in chapter 3.

5. One-photon spectroscopy of the $(1)\,^3\Sigma_g^+$ potential

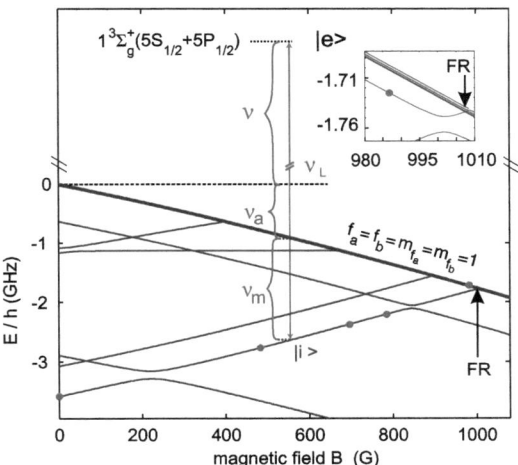

Figure 5.2: Magnetic shifts of molecular levels in the electronic ground state. As we refer all energies to the $(5S_{1/2} + 5S_{1/2})$ asymptote at 0 G, the scheme shows how we calculate the magnetic shifts of the excited levels. The graph shows a simplified molecular level scheme of $^{87}$Rb$_2$ in the $a\,^3\Sigma_u^+$ potential. For an improved visibility we only show levels with rotational quantum number $l = 0$ (s-waves). As energy reference ($E = 0$) we take the dissociation threshold of two atoms in their electronic ground states (that is $f = 1$, $m_f = 1$) at 0 G. The laser light couples the molecular level $|i\rangle$ and the excited level $|e\rangle$ with frequency $\nu_L$. It can be tuned to almost any deeply bound level of the $(1)\,^3\Sigma_g^+$ potential. The red dots indicate the states where we perform our measurements. Obviously the transition energy depends on the magnetic field. For simplicity, the magnetic shift in the excited state is not shown. The inset displays the region near the Feshbach resonance at 1007.4 G (indicated with an arrow and "FR").

57

# 5. One-photon spectroscopy of the $(1)\,^3\Sigma_g^+$ potential

the number of molecules via a reverse Feshbach magnetic field sweep, dissociating the remaining $|i\rangle$ molecules into atoms which are then detected by absorption imaging.

After we have produced the Feshbach molecules, the magnetic field is normally set to 986.8 G where the spectroscopy is carried out. In Fig. 5.2, we show the magnetic structure of molecules over the whole range of 1000 G which we use in the present experiments to extract the shifts of the excited states. For simplicity, we only display molecular levels with $l = 0$. A detailed discussion of higher rotational levels can be found in [Lang 2008a]. At 986.8 G the binding energy of the Feshbach molecules is 22.7 MHz×$h$. This is depicted via red dots in Fig. 5.2. Their state vectors can be approximated by Hund's coupling case e) with atomic quantum numbers

$$|v, (f_a, f_b)f, F, M_F\rangle.$$

Here $v$ corresponds to the vibrational quantum number and $f_a$, $f_b$ are the total angular momenta for atoms $a$ and $b$, respectively. This nomenclature is in accordance with chapter 4. The sum of both atomic angular momentum operators $\mathbf{f} = \mathbf{f}_a + \mathbf{f}_b$ couples with the operator for pure molecular rotation $\mathbf{l}$ to form the total angular momentum $\mathbf{F} = \mathbf{f} + \mathbf{l}$. At high magnetic fields the total angular momentum $\mathbf{F}$ is no longer a good quantum number but its projection $M_F$ onto a space fixed axis remains good. According to this basis the prepared Feshbach state $|i\rangle$ will be approximated by the state vector $|v = 40, (f_a = 2, f_b = 2)f = 2, l = 0, F = 2, M_F = 2, \rangle$ at low fields. At 986.8 G, $f_a$ and $f_b$ become bad quantum numbers and the expectation values[2] are about 1.5.

For magnetic fields lower than 986.8 G we use a molecular level for the Feshbach molecules which either correlates to $|v = 36, (f_a = 2, f_b = 2)f = 2, N = 0, F = 2, M_F = 2\rangle$ or $|v = 40, (f_a = 1, f_b = 1)f = 2, N = 0, F = 2, M_F = 2\rangle$ at low magnetic fields. The state with $v = 36$ is the diagonal line in Fig. 5.2 going from the point $(B = 0, E/h = -3.6$ GHz) to the position of the Feshbach resonance at threshold. This level exhibits several avoided crossings with other molecular levels. When ramping down the magnetic field to its specified value these avoided crossings have to be crossed. For this we employ an adiabatic transfer method at the avoided crossings on the basis of radiofrequency transitions as described in [Lang 2008a]. In order to count the remaining molecules after the spectroscopy pulse we retrace the path back to the Feshbach resonance at 1007.4 G where the molecules dissociate.

To calculate magnetic shifts and excitation energies we refer all energies to two atoms in their ground states $|f = 1, m_f = 1\rangle$ at 0 G. For this, we subtract from the measured laser excitation energy $h\nu_L$ both the bound state energy of the Feshbach level $h\nu_m$ as well as the Zeeman energy $h\nu_a$ of the free atoms (see Fig. 5.2), which are both well known (see [Strauss 2010]).

For the later discussion of our spectra and the lines which we can access, it will be also important to know the symmetries of our Feshbach state. This was described in

---

[2] We define expectation values for the quantum numbers of an operator as $\sum_i \lambda_i |\langle e_i | \Psi \rangle|^2$, where $\lambda_i$ are the eigenvalues of the operator, $e_i$ are corresponding normalized basis vectors, and $|\Psi\rangle$ is the state under consideration.

chapter 4 and I will only summarize the results. The Feshbach state has positive total parity with respect to inversion of both the electronic and nuclear coordinates since the parity is given by $(-1)^N$, and the molecules are in the rotational ground state of the $a\,^3\Sigma_u^+$ potential. Furthermore, one can show that owing to the antisymmetry of the molecular wave function with respect to nuclear exchange (nuclear spin of $^{87}$Rb $i = \frac{3}{2}$), molecules with even $N$ in the $a\,^3\Sigma_u^+$ state must have "+" symmetry and either a total nuclear spin $I = 1$ or 3 [Herzberg 1950, Townes/Schawlow 1955]. The Feshbach state also has contributions from singlet states and the expectation values for the total nuclear spin and the electron spin are $I \approx 1.7$ and $S \approx 0.8$, respectively. This is due to the fact that a superposition of the $X\,^1\Sigma_g^+$ and $a\,^3\Sigma_u^+$ states does not have a proper "g/u" symmetry anymore. For further details on the $a\,^3\Sigma_u^+$ ground state see chapter 6 and [Strauss 2010]. Since here we are interested in transitions to the $(1)\,^3\Sigma_g^+$ potential, only the triplet contributions in the Feshbach molecules are relevant.

To further investigate the magnetic moments and to check the level assignment, we also applied the method of **A**diabatic **T**ransfers across **A**voided **C**rossings (ATAC) to go to lower magnetic fields [Lang 2008a]. This method is necessary to circumvent uncontrolled losses when sweeping over an avoided crossing with the magnetic field and enables us to obtain experimental data on the Zeeman shift of the excited state. (See Fig. 5.2 for the magnetic structure.) Therefore, we insert transfers between the Feshbach association and the laser pulse using radio frequency. I will briefly give the principle of an ATAC which is applied several times: When the Feshbach molecules are close to an avoided crossing we switch on a radio frequency field to couple the upper and lower branches of the avoided crossing. Lowering the magnetic field adiabatically further, we reach the other branch of the avoided crossing. Switching off the radio frequency field completes the transfer of the molecules to the lower branch. To detect the molecules we apply the reverse sequence after the spectroscopy laser is switched off.

### 5.1.1 Stability of lasers and uncertainties

For spectroscopy we either use a free-running Ti:Sapphire laser or a grating-stabilized diode laser. Because the unlocked lasers typically drift a few MHz during one experimental cycle, we have the option to lock them to a cavity using the Pound-Drever-Hall scheme. This was mainly utilized for a few weak lines which are located close to broad lines (see Fig. 5.6). A high precision of less than 1 MHz and high short term stability are mandatory. The cavity is in turn locked to an atomic $^{87}$Rb line. If we lock the laser to this cavity it has a short term laser line-width of less than 100 kHz and its beam has a $1/e^2$ intensity waist radius of 130 μm at the molecular sample. The light is polarized parallel to the magnetic bias field $B$ (pointing parallel to the gravitational field in the z-direction) and thus can only induce $\pi$ transitions. The frequency is read out automatically using a commercial wave meter (WS7 from HighFinesse). It has a nominal accuracy of 60 MHz after calibration. Over several days we have observed drifts of $\pm 200$ MHz, for example by repeatedly addressing the same excited state. Over the length of a few

# 5. One-photon spectroscopy of the $(1)\,^3\Sigma_g^+$ potential

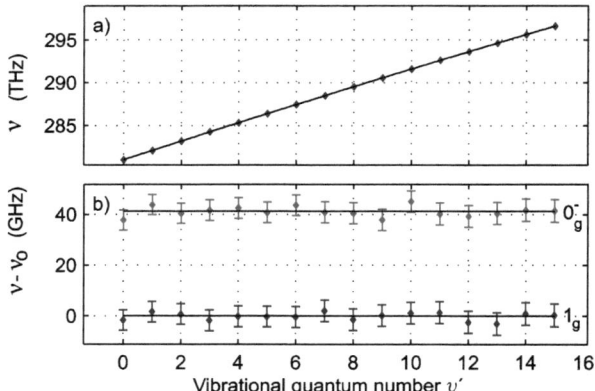

Figure 5.3: Vibrational ladder for the $(1)\,^3\Sigma_g^+$ potential. a) Excitation frequencies $\nu$ of the $1_g$ states. The continuous line is a second order fit to the data according to a Morse potential. The Fit gives $a = (0.271 \pm 0.004)/a_0$ and $D = (3240 \pm 50)\,\text{cm}^{-1}$ b) Vibrational progression of the $0_g^- - 1_g$ splitting which we measure at high power and low resolution. Here, the offset $\nu_0$ is a function of the vibrational quantum number $v'$. The lower branch corresponds to the $1_g$ part and the upper one to $0_g^-$. For each vibrational quantum number we take the excitation frequency $\nu$ of the $1_g$ fit as frequency reference $\nu_0(v') = 0$. As the position of each line we take the center of each spectrum at high power. Note the different energy scale of a) and b).

experimental cycles (5 minutes) the wave meter is stable to within 10 MHz, which represents a random noise floor. To increase the reliability of the wave meter we calibrate it every day using the atomic $|f = 1\rangle \to |f' = 2\rangle$ transition of the repumper laser which we use for the magneto-optical trap. This laser is locked to an atomic $^{87}$Rb line using frequency modulation spectroscopy. Based on our experience with the wave meter where we have measured excitation energies of a few lines over an extended period of time we estimate an experimental accuracy of about 60 MHz. This includes peak position uncertainties due to number fluctuation as well as a frequency drift of the laser between the laser pulse and wavelength measurement.

## 5.2 Experimental observations

In a first set of experiments, we have mapped out the vibrational ladder of the $(1)\,^3\Sigma_g^+$ potential from $v = 0$ to $v = 15$ at low resolution (see Fig. 5.3 a)). We used the Ti-sapphire laser with maximum available power of a few 100 mW such that we only observed broad lines with a typical width of several GHz. The magnetic field was set to 986.8 G. The vibrational ground state of the $(1)\,^3\Sigma_g^+$ potential has an excitation frequency of 281.1 THz with respect to $|i\rangle$, corresponding to a laser wavelength of 1067nm. We checked for potentially deeper bound states by searching at even higher wavelengths but nothing was found. The vibrational splitting between the two lowest levels, $v = 0$ and $v = 1$, is about 1.2 THz. The solid line in Fig. 5.3 a) corresponds to a quadratic fit to the data which leads to the same Morse potential as given by Mudrich el. al. [Mudrich 2010]. Fig. 5.4 shows the same data together with the corresponding Born-Oppenheimer and Morse potentials. It can be seen nicely that we only measured the bottom of the potential. The inset shows a zoom into the region where we took data. I show the measured vibrational levels together with the energies calculated from the Born-Oppenheimer potential (dotted line). The deviations from the fit are tiny and difficult to see. I want to mention that the absolute energy with respect to the $5S_{1/2} + 5S_{1/2}$ asymptote of the Morse potential is only determined experimentally.

In order to resolve the whole rotational, hyperfine, and Zeeman structures, we reduce the power. We then observe that each vibrational state splits into two parts, which are 47 GHz[3] apart as we show in Fig. 5.3 b). In this plot we take the center of each $0_g^-/1_g$ vibrational line. For the nomenclature we use a Hund's case c) notation $|\Omega|_g$ with the projection $\Omega$ of $J$ onto the internuclear axis. For our $^3\Sigma$ state $J$ is composed of $R$ and the total electronic spin $S$. In our case $\Omega = \Lambda + \Sigma = \Sigma$ holds, where $\Lambda$ ($\Sigma$) and $\Omega$ are the projections of the electronic angular momentum $L$ (spin $S$) and the total angular momentum $J = N + S$ on the internuclear axis, respectively. The large error bars of several GHz reflect the crudeness of this first measurement where we do not resolve the substructure of each of the doublet components.

### 5.2.1 The splitting of the vibrational levels into $0_g^-$ and $1_g$ components

The large splitting of the vibrational levels by 47 GHz clearly cannot be explained by the rotational, hyperfine or Zeeman interactions. The rotational constant $B_v$ should be around 400 MHz×$h$ for the low lying vibrational levels in $(1)\,^3\Sigma_g^+$ potential. Estimating the hyperfine and Zeeman energies from those for $^{87}$Rb atoms we expect such contributions to be at most a few GHz. It turns out that the splitting stems from a strong effective spin-spin coupling of the electrons. Besides direct spin-spin interaction of the electrons second-order spin-orbit coupling also contributes to this coupling. In fact, second order spin-orbit coupling should be dominant as it is resonantly enhanced by a $(1)^1\Pi_g$ state nearby. (See Fig. 5.4 with the potentials from [Lozeille 2006] and [Kayama 1967].) Experimentally these two contributions cannot be separated and the spin-spin interaction takes the form (See section 5.3.1)

$$H_{\text{ss}} = 2\lambda \mathbf{S}_Z^2. \tag{5.1}$$

---

[3]Here we include the rotational substructure. If we ignore it, we obtain a slightly lower value of $2\lambda = 42$ GHz.

## 5. One-photon spectroscopy of the $(1)\,^3\Sigma_g^+$ potential

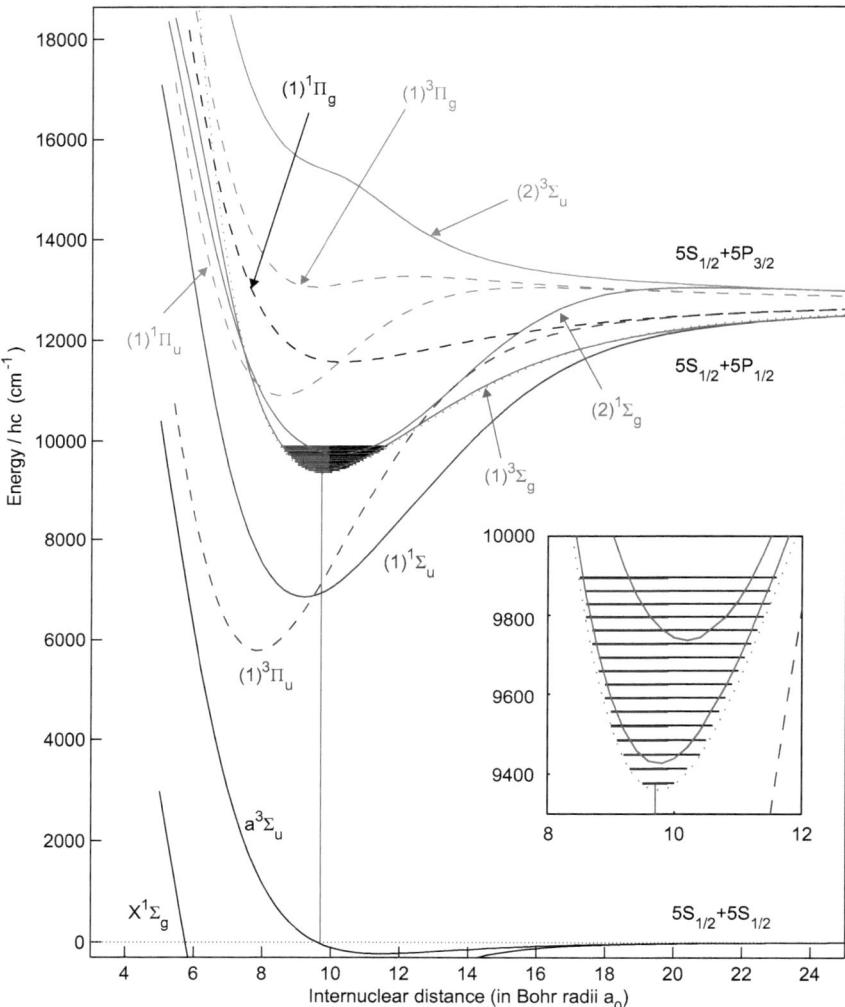

Figure 5.4: Hund's case a) potential structure of the first excited $(1)\,^3\Sigma_g^+$ state of $^{87}$Rb$_2$. The inset shows a zoom into the region where the spectroscopy was carried out. Calculations from [Lozeille 2006]. For details see text.

A strong $H_{ss}$ couples the electronic spin to the internuclear axis, making its projection $\Sigma = 0, 1$ a good quantum number. Thus, for our $(1)\,^3\Sigma_g^+$ state $\Omega = \Sigma$ is also a good quantum number. The energy eigenvalues of $H_{ss}$ are $2\lambda \cdot \Sigma^2$, which means that the splitting is simply $2\lambda$. Our data then indicate a $2\lambda \approx 47\,\text{GHz}$ as mentioned above. Such a large $\lambda$ leads to the dominant doublet structure where the more deeply bound doublet component has $1_g$ character and the other one has $0_g^-$ character (see Fig. 5.3 b)). From a different point of view we can interpret all the lines as belonging to their respective $1_g$ / $0_g^-$ Hunds case c) potentials which are shifted by $\approx 47\,\text{GHz}$ with respect to each other. The two vibrational ladders are parallel indicating that the potentials have the same shape. The final analysis in sections 5.4.1 and 5.4.2 shows that the rotational constants are similar which means that the equilibrium internuclear separation $R_e$ also coincides. As a result we can regard these states to a good approximation as belonging to the $(1)\,^3\Sigma_g^+$ electronic state.

As we lower the power further we observe a rich substructure which was not entirely understood in the $1_g$ case. This structure is spread out over $3\,\text{GHz}$ ($12\,\text{GHz}$) in the case of $0_g^-$ ($1_g$). As an example, we show the overall rotational and hyperfine structure in Fig. 5.5 for the $1_g$ $v = 0$ and 13 vibrational states. In this figure, the most deeply bound line within the hyperfine and rotational structure lies at $0\,\text{GHz}$. The y-axis shows the molecular fraction remaining. Furthermore, one can see that the spectra for different deeply bound vibrational levels look similar, as expected. We exclude the possibility that the substructure is accidentally caused by states from other Born-Oppenheimer potentials. These spectra typically consist of roughly 300 points where each point corresponds to one production and measurement cycle which takes $28\,\text{s}$. We observe some ten lines in each spectrum, which vary markedly in linewidth. The width of each line is determined by the coupling between the levels $|i\rangle$ and $|e\rangle$ and the strength of the light field, that is, the Rabi frequency $\Omega$.

### 5.2.2 The spectra of the $0_g^-$ and $1_g$ states

The $0_g^-$ and $1_g$ states have a rich substructure which we are able to resolve by lowering the power of the laser to about $0.1\,\text{mW}$. Figure 5.5 shows loss spectra for $v' = 0$ and $v' = 13$, respectively. It is hard to recognize a pattern among the 12 observed $1_g$ lines, resulting from an interplay of rotation, hyperfine interaction and Zeeman interaction. It is one of the main goals of this chapter to understand this spectral pattern and to identify the individual lines.

The varying linewidth indicates a strong variation of the laser-induced coupling between the different $|e\rangle$ and $|i\rangle$ levels. In order to make sure that all of these lines belong to the $(1)\,^3\Sigma_g^+$ potential and not to some other overlapping potentials ($(1)\,^3\Pi_u$ or $(1)\,^1\Sigma_u^+$) we have taken spectra for different deeply bound vibrational levels. The $v' = 0$ and $v' = 13$ spectra a) and b) in Fig. 5.5, for example, are clearly similar. Compared to the typical step size of $40\,\text{MHz}$ in Fig. 5.5, the natural linewidth of the molecular levels of $12\,\text{MHz}$ is relatively small. Thus, it is possible that some weak lines are not always detected

# 5. One-photon spectroscopy of the (1) $^3\Sigma_g^+$ potential

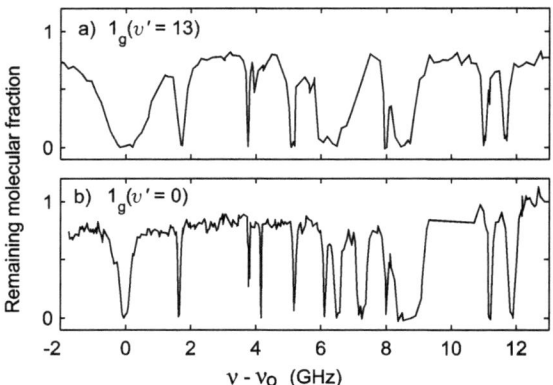

Figure 5.5: The spectra of two different $1_g$ vibrational manifolds of the (1) $^3\Sigma_g^+$ state at 986.8 G. For each spectrum we subtract a frequency offset of $c$ from the excitation frequency $\nu$ such that the most deeply bound line is situated at 0 GHz. a) The $v = 13$ spectrum with an offset of $\nu_0 = 294624.4$ GHz. b) The vibrational ground state where we subtract an offset of $\nu_0 = 281066.2$ GHz.

in a single scan, especially when they are located on the shoulder of a strong line. By testing for consistency, and by checking theoretical predictions we gradually completed the search for lines. For the $1_g$ spectra we found in total 18 lines which we present in our discussion in section 5.4.2.

While the $1_g$ spectrum is spread out over 12 GHz, the $0_g^-$ manifold has a much smaller spread of 3 GHz and fewer lines (Fig. 5.6). We observed 5 lines which are arranged in a doublet like structure. The doublet splitting is $\approx 2.5$ GHz between the single line at 0 GHz and the other 4 lines. This splitting can be understood in terms of molecular rotation **R**,

$$H_\text{rot} = B_{v'} \cdot \mathbf{R}^2/\hbar^2, \tag{5.2}$$

where $B_{v'}$ is the rotational constant and $\mathbf{R}/\hbar^2$ is the rotational angular momentum of the two nuclei in units of $\hbar$. In comparison, the contributions of hyperfine and Zeeman interactions are much weaker here. For the $0_g^-$ state, the Zeeman interaction and the hyperfine interaction have diagonal matrix elements proportional to $\Omega$, and hence their contribution is off-diagonal and rather small (section 5.3). The rotational constant $B_{v'}$ is computed from the vibrational wave function $\psi_{v'}(R) = \langle R | v' \rangle$

$$B_{v'} = \frac{\hbar}{4\pi\mu} \left\langle v' \left| \frac{1}{R^2} \right| v' \right\rangle \tag{5.3}$$

where $|v'\rangle$ is the ket for the vibrational wave function and $\mu$ is the reduced mass. From our Morse potential and recent *ab initio* calculations [Lozeille 2006] we expect a value

5. One-photon spectroscopy of the $(1)\,^3\Sigma_g^+$ potential

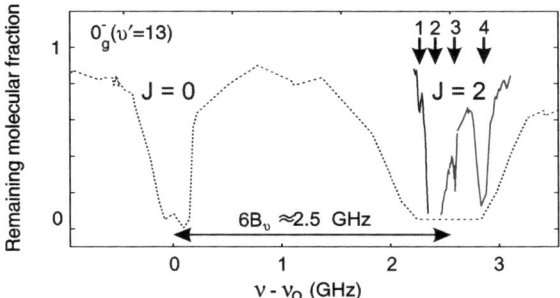

Figure 5.6: High-resolution scan of $0_g^-(v'=13)$ lines in the $(1)\,^3\Sigma_g^+$ potential at a magnetic field of 986.8 G. Here, scans taken at various laser powers and pulse lengths are merged into one graph. The dotted line corresponds to a laser power of 0.1 mW and a pulse duration of 200 ms while the continuous lines with indices 1, 2, 3, 4 were measured with a laser power of 0.1 mW and 1 ms pulse duration. The frequency offset from the excitation frequency $\nu$ is $\nu_0 = 294671.0$ GHz.

of about $B_{v'=13} = 400$ MHz. Due to the weak hyperfine and Zeeman interactions the angular momentum $J$ ($\mathbf{J} = \mathbf{S} + \mathbf{L} + \mathbf{R}$) is a good quantum number here. The rotational splitting is then determined by

$$E_{\rm rot} = B_{v'}\, J(J+1). \tag{5.4}$$

Here, we have to use Eq. (5.2) and the definition of $\mathbf{J}$ to obtain the nontrivial result above. (See for example [Wilson 1955, Lefebvre-Brion/Field 2004].) In our spectrum we observe the line with $J = 0$ at low excitation energy while the line with $J = 2$ is shifted up by $6B_{v'} \approx 2.5$ GHz. The $J = 1$ rotational level is not accessible because total parity (inversion of electron and nucleon coordinates) has to change in the optical transition. This result is obvious, when looking at section 4.7.

In contrast to the $J = 0$ line, the $J = 2$ line has a substructure of four lines which arises from residual hyperfine, and Zeeman interactions. This substructure will be discussed in detail in section 5.4 along with the analysis of our theoretical model which we use to describe the full substructure within one vibrational manifold. In order to obtain the spectrum, various scans at different laser power were carried out which were then merged into a single spectrum. These weak lines were only found after a search which was motivated in order to check the predictions of our theoretical model.

## 5.3 Effective Hamiltonian and evaluation of molecular parameters

In the following we present the diatomic Hamiltonian which we use to explain the observed energy levels within the deeply bound $v' = 13$ vibrational manifold of the $(1)\,^3\Sigma_g^+$ potential. Line spectra in other low-lying vibrational manifolds are similar (Fig. 5.5) and are essentially described by the same Hamiltonian with only slightly adjusted parameters. The Hamiltonian reads

$$H = H_{\text{ss}} + H_{\text{rot}} + H_{\text{hf}} + H_{\text{Z}} + H_{\text{sr}}. \tag{5.5}$$

$H_{\text{ss}} = 2\lambda \mathbf{S}_Z^2$ is the effective spin-spin operator as given in Eq. (5.1) which, as previously discussed, leads to the large splitting into the $0_g^-$ and $1_g$ components. $H_{\text{rot}} = B_{v'} \cdot \mathbf{R}^2$ is the Hamiltonian for molecular rotation as already discussed in Eq. (5.2). The spin-spin interaction $H_{\text{ss}}$, hyperfine interaction $H_{\text{hf}}$, Zeeman interaction $H_{\text{Z}}$, and finally spin-rotational interaction $H_{\text{sr}}$ are explained in detail in this section. Since it is the goal of this chapter to get a first understanding of the experimentally observed spectra, we will in general simplify the interaction and only take into account terms of leading order [Veseth 1976, Veseth 1976a]. For the derivation of the effective Hamiltonian and for analytical expressions of the matrix elements we refer the reader to [Lysebo 2009][4].

### 5.3.1 Direct spin-spin and second-order spin-orbit interaction

It is well known that the direct spin-spin and second order spin-orbit interaction have an effective Hamiltonian of the form

$$H_{SS} + H_{SO} = \frac{2}{3}(\lambda_{SS}(R) + \lambda_{SO}(R)) \cdot \left(3 S_Z^2 - \mathbf{S}^2\right), \tag{5.6}$$

where we use the internuclear axis as the $Z$-axis and the internuclear separation $R$ [Lefebvre-Brion/Field 2004, Brown/Carrington 2003]. These Hamiltonians are used in the work of [Mies 1996] to give analytical approximations for the $R$-dependence of $\lambda$. As I will explain in chapter 6, we use the expression from Mies et al. including the $R$ dependence in our coupled-channel calculation for the ground state. For the excited state, the $R$-dependence of $\lambda$ does not appear explicitly because we use an effective Hamiltonian within one vibrational level. Although Eq. (5.6) is used frequently, it is difficult to find the reason for this functional dependence explained in the literature, especially in the case of second-order spin-orbit coupling. A first account on the effective spin-spin Hamiltonian is given by Kramers [Kramers 1929, Kramers 1929a], who used group theory to obtain Eq. (5.6). Here, I will follow the approach of Tinkham [Tinkham 1954], who treats first- and second-order effects in the fine structure using wave mechanics.

---

[4]Here, one has to account for the different axes of quantization. In the analysis one then has to take the anomalous commutation relations for angular momenta which involve the rotation of the nuclei [Brown 1976, VanVleck 1951]

**Spin-spin contribution and $\lambda_{SS}$**

We start with the basic microscopic Hamiltonian [Tinkham 1954] of the form

$$H_{SS} = C \cdot \sum_{j,k} \frac{(\mathbf{s}_j \cdot \mathbf{s}_k)r_{jk}^2 - 3(\mathbf{s}_j \cdot \mathbf{r}_{jk})(\mathbf{s}_k \cdot \mathbf{r}_{jk})}{r_{jk}^5}, \qquad (5.7)$$

where $\mathbf{r}_{jk} = \mathbf{r}_j - \mathbf{r}_k$ and $C = g_S^2 \mu_B^2 (\mu_0/4\pi)$. Here $g_S$ is the electronic g-factor, $\mu_B$ is the electron Bohr-magneton, and $\mu_0$ is the magnetic permeability. Expanding $H_{SS}$ into Cartesian components and regrouping leads to

$$\begin{aligned} H_{SS} = &-C \cdot \sum_{j,k} \frac{x_{jk} y_{jk}}{r_{jk}^5} (s_{jx} s_{ky} + s_{jy} s_{kx}) \\ &+ \frac{3 y_{jk} z_{jk}}{r_{jk}^5} (s_{jy} s_{kz} + s_{jz} s_{ky}) + \frac{3 z_{jk} x_{jk}}{r_{jk}^5} (s_{jz} s_{kx} + s_{jx} s_{kz}) \\ &+ \frac{3}{2} \frac{x_{jk}^2 - y_{jk}^2}{r_{jk}^5} (s_{jx} s_{kx} - s_{jy} s_{ky}) + \frac{1}{2} \frac{3 z_{jk}^2 - r_{jk}^2}{r_{jk}^5} (2 s_{jz} s_{kz} - s_{jx} s_{kx} - s_{jy} s_{ky}), \end{aligned} \qquad (5.8)$$

where $j, k$ refer to the different electrons and $x, y, z$ stand for the different axis. Now we integrate Eq. (5.8) over all electronic coordinates to obtain the effective Hamiltonian. In this step the first three terms vanish because they involve a product of an odd function like $x_{jk}$ and an even function $1/r_{jk}^5$. The fourth term vanishes because of the axial symmetry of a diatomic molecule, that is the integral over $x_{jk}^2$ is the same as for $y_{jk}^2$. Thus, only the last term remains. This term is proportional to $(3S_Z^2 - \mathbf{S}^2)$ [VanVleck 1951]. I refer the reader to [Tinkham 1954] for the actual *ab initio* estimate of $\lambda_{SS}$, which is calculated via the integration mentioned above. A rigorous derivation of the spin-spin interaction using irreducible tensor operators can be found in [Brown/Carrington 2003], page 563.

**Second-order spin-orbit contribution and $\lambda_{SO}$**

The energy contribution of the second-order spin-orbit interaction is [Marinescu 1995]

$$H_{SO}^{(2)} = -\sum_n \frac{\langle n \mid A\mathbf{L} \cdot \mathbf{S} \mid 0 \rangle \langle n \mid A\mathbf{L} \cdot \mathbf{S} \mid 0 \rangle^*}{E_n - E_0}, \qquad (5.9)$$

where the $E_n$ refer to the unperturbed energies that is without fine structure. This is standard perturbation theory where we denote the excited orbital state by $|n\rangle$, and the coupling constant for spin-orbit interaction by $A$. The term containing $L_Z S_Z$ is diagonal and is not of interest at the moment as we are dealing with $\Sigma$ states. As the orbital wave functions do not act on the spin, the matrix elements can be written $\sum_g \langle n \mid AL_g \mid 0 \rangle S_g$, where $g = X, Y$. Standard angular momentum theory shows that the only non-zero

# 5. One-photon spectroscopy of the (1) $^3\Sigma_g^+$ potential

matrix elements are $\langle \Lambda | L_{X,\,Y} | \Lambda \pm 1 \rangle$ and we have $n = 1$ [VanVleck 1951]. Thus, we can simplify the Hamiltonian to

$$H_{SO}^{(2)} = -\sum_{n=1}\sum_{g,\,g'} \frac{\langle n | AL_g | 0 \rangle S_g \langle n | AL_{g'} | 0 \rangle^* S_{g'}^*}{E_n - E_0}.$$

Here, the superscript (2) is a reminder that we are dealing with a second order effect. A careful analysis of angular momentum operators in space-fixed and molecule-fixed systems shows that [VanVleck 1951]

$$\langle \Lambda | L_Y | \Lambda \pm 1 \rangle = \pm i \langle \Lambda | L_X | \Lambda \pm 1 \rangle. \tag{5.10}$$

Eq. (5.10) implies that all cross terms of the form $(g) \cdot (g')$ drop out in Eq. (5.9) and we have

$$H_{SO}^{(2)} = \sum_{n=1} \frac{|\langle n | AL_X | 0 \rangle|^2}{E_n - E_0} \left( S_X^2 + S_Y^2 \right).$$

If we now use the fact that we are interested in second order effects, we have to interpret the last factor in terms of a tensor operator of rank 2. As already mentioned, the $Z$ component of $\mathbf{L}$ corresponds to diagonal elements, which we ignore. Altogether, we need to find the $m = 0$ component of a tensor operator of rank two:

$$\begin{aligned}\left( S_X^2 + S_Y^2 \right) &= \sqrt{6}\, T_{m=0}^{(2)}(\mathbf{S}, \mathbf{S}) \\ &= 2S_Z^2 - S_X^2 - S_Y^2.\end{aligned}$$

These steps involve standard tensor operator algebra and can for example be found for the present problem in [Brown/Carrington 2003]. This completes the proof and we finally obtain

$$\begin{aligned}H_{SO}^{(2)} &= \frac{2}{3}\lambda \left( 3S_Z^2 - \mathbf{S}^2 \right) \\ \lambda &= \sum_{n=1} \frac{|\langle n | AL_X | 0 \rangle|^2}{E_n - E_0}.\end{aligned} \tag{5.11}$$

Theoreticians can use Eq. (5.11) to estimate the strength of this interaction with the help of the nearby $\Pi$ states.

### 5.3.2 The hyperfine interaction

The second crucial part in the Hamiltonian is the hyperfine-structure, as the atomic hyperfine splitting is about $7\,\text{GHz} \times h$. Here, the first theoretical work for diatomic molecules goes back to the 1950s, when an effective Hamiltonian in a Hund's coupling scheme $a_\beta$) and $b_\beta$) was derived [Frosch 1952]. Afterwards the theory was extended to Hund's coupling case c) and includes the work of [Mustelin 1963, Freed 1966] and

[Veseth 1976, Kristiansen 1986]. These articles also extend the theory to interactions between the gradient of the electric field and the quadrupole field of the nucleus. Experiments have been carried out using for example $J_2$ to observe the quadrupole hyperfine structure [Hänsch 1971]. Latest results in the field of ultracold molecules were obtained with KRb [Ospelkaus 2010] and $Cs_2$ [Danzl 2010].

A general ansatz for the hyperfine Hamiltonian in the Hund's case a) and b) is of the form [Townes/Schawlow 1955]

$$H_{\text{hf}} = a\Lambda I_Z + (b_F - \frac{1}{3}c) \cdot \mathbf{I} \cdot \mathbf{S} + cI_Z \cdot S_Z. \tag{5.12}$$

Here, $\mathbf{S}$ ($\mathbf{I}$) is the operator for the total electron (nuclear) spin and $Z$ denotes the internuclear axis. The first term describes the interaction of the electronic angular momentum with the nuclear spin. However, since $\Lambda = 0$ this term will not contribute. The second and third terms include the Fermi contact term and the so-called anisotropic interaction. $b_F$ is the Fermi contact parameter while $c$ is called the anisotropic hyperfine parameter. For $\Sigma$ states we have $c \ll b_F$ ([Townes/Schawlow 1955], page 196). Marius Lysebo and Leif Veseth tried to calculate the hyperfine parameters *ab initio* but were not able to obtain reliable results. Therefore, they have used these parameters in the fit.

### 5.3.3 The Zeeman interaction

As we carry out our measurements at magnetic fields of up to 1000 G, Zeeman interaction plays an important role. The main contribution to the Zeeman interaction comes from the electrons while contributions from the nuclear spins and molecular rotation are much smaller and are neglected here (as well as other second-order effects treated in [Veseth 1976a]). In general, the Zeeman interaction is

$$H_Z = \mu_B g_S S_z \cdot B_z + \mu_0 \left(g_I I_z + g_N N_z\right) B_z, \tag{5.13}$$

where $B_z$ is the magnetic field pointing in the space-fixed $z$-direction, $\mu_B$ is the Bohr magneton, and $g_S = 2$ is the electron g-factor. Since $\Lambda$ vanishes in the $(1)\,^3\Sigma_g^+$ state, the third term due to the orbital angular momentum of the electron is expected to be small. The same holds for the nuclear part involving $I_z$. We note that there is no free parameter in the Zeeman Hamiltonian to adjust the model to the measured data. While the contribution from the Zeeman interaction will in general be large for the $1_g$ lines, it will be small for the $0_g^-$ lines because here the spin projection on the internuclear axis is $\Sigma = 0$.

### 5.3.4 The rotational interaction

The simplified energies in Eq. (5.4) are an approximation where we already include the total electronic spin $S$. In the code for the numerical calculations which was used by our theoreticians Marius Lysebo and Leif Veseth, the rotational Hamiltonian is

$$H_{\text{rot}} = B_v \mathbf{R}^2 \tag{5.14}$$

with $B_v$ the rotational constant, and $\mathbf{R}$ the orbital angular momentum of the relative motion of the two nuclei, which is in the case of $\Sigma$ states $\mathbf{R} = \mathbf{J} - \mathbf{S}$. This operator is in general only diagonal in a basis with quantum numbers $S, I_1, I_2, F, M_F$, but off-diagonal in $\Omega, \Omega_{I_1}$ and $\Omega_{I_2}$. As already mentioned, we obtain the value $B_{v'=13} = 412\,\text{MHz}$. We also included the rotational constant in the fit routine but did not observe any noticeable change in the value. The centrifugal distortion constant $D$ is significantly smaller than $B_v$. In fact, $\frac{D}{B_v} \simeq 10^{-6}$, which means that this effect may safely be neglected in the present work.

### 5.3.5 The spin-rotational interaction

In principle, we can also include the spin-rotational interaction through the effective Hamiltonian

$$H_{\text{sr}} = \gamma_{v'} \mathbf{N} \cdot \mathbf{S}. \tag{5.15}$$

where $\mathbf{N} = \mathbf{L} + \mathbf{R}$ is the operator for molecular rotation including the total orbital angular momentum and $\gamma_{v'}$ is the spin-rotation coupling constant. However, $\gamma_{v'}$ is typically a small fraction of the rotational constant $B_{v'}$ [Lefebvre-Brion/Field 2004] and only represents an insignificant correction to the energy levels in the present system and is thus neglected.

### 5.3.6 Fit procedure and evaluation of molecular parameters

According to the Hamiltonian in Eq. (5.5), there are four adjustable parameters (neglecting $\gamma$): the rotational constant $B_{v'}$, the spin-spin splitting parameter $\lambda$, the Fermi-contact parameter $b_F$, and the anisotropic hyperfine parameter $c$. As discussed in section 5.2.2, the rotational constant $B_{v'}$ should be close to 412 MHz from *ab initio* calculations for the $(1)\,^3\Sigma_g^+$ potential and as indicated from the $1_g$ spectrum in Fig. 5. The spin-spin parameter $\lambda$ is mainly determined by the 47 GHz splitting of the $1_g$ and $0_g^-$ manifolds. This only leaves $b_F$ and $c$ as completely free parameters. Marius Lysebo and Leif Veseth determine the parameters from fits of the model to the experimental data using a nonlinear Levenberg-Marquardt method [Marquardt 1963, Press *et al.* 2007]. For the calculations, the Hamiltonian in Eq. (5.5) is expressed in terms of matrix elements in a Hund's case $a_\alpha$) basis,

$$|\Lambda, S, \Sigma, I_1, I_2, \Omega_{I_1}, \Omega_{I_2}, F, \Omega_F, M_F\rangle, \tag{5.16}$$

where $\Omega_F = \Lambda + \Sigma + \Omega_{I_1} + \Omega_{I_2}$ is the projection of the total spin on the internuclear axis. $I_1, I_2$ are the spins of the two nuclei and $\Omega_{I_1}, \Omega_{I_2}$ are their projections on the internuclear axis. For the analytical expressions of the matrix elements as a function of the quantum numbers of Eq. (5.16) I refer the reader to [Lysebo 2009]. After the evaluation of the Hamiltonian in the basis above, our theoreticians obtain a matrix which is diagonalized numerically to arrive at the eigenvalues and eigenstates. Included in this calculation are all hyperfine states in the $^3\Sigma_g^+(v'=13)$ electronic state with total angular momentum up to $F=10$. Experimentally, we only observe states with $F<7$. The best fit results in the following values for the fit parameters: $B_{v'}=412\,\text{MHz}$ and $\lambda=47\,\text{GHz}$. The parameters $b_F$ and $c$ are determined in terms of the combinations $b_F+\frac{2}{3}c$ and $b_F-\frac{1}{3}c$, which correspond to contributions diagonal and off-diagonal in $\Sigma$ and $\Omega$. We find $b_F+\frac{2}{3}c=832\,\text{MHz}$. In contrast, we find that $b_F-\frac{1}{3}c$ is not precisely determined in our analysis. For further details see section 5.4.1.

## 5.4 Discussion of results

As already discussed in section 5.1, we can obtain a rough understanding of the observed substructure by looking at the rotational and spin-spin Hamiltonian. Now we also want to include the hyperfine and magnetic interactions. Our initial Feshbach state $|i\rangle$ does not have well defined quantum numbers $S$ and $I$ (section 5.1). This is in contrast to the excited $(1)\,^3\Sigma_g^+$ state where we have $S=1$ and $M_F=2$ (and thus $F\geq 2$) for all states. In the following discussion we will omit the quantum numbers $S$ and $M_F$. For $2\lambda=47\,\text{GHz}$, the interactions between the $\Omega=0$ and $\Omega=1$ substates are small, and energy levels are robust against small variations in $\lambda$.

### 5.4.1 $0_g^-$ spectrum and magnetic shifts

The $0_g^-$ spectrum is significantly simpler than the $1_g$ spectrum. Computed energy levels as a function of the magnetic field strength are shown in Fig. 5.7, along with experimental data. The x-axis displays the magnetic field strength in Gauss, the y-axis corresponds (up to an offset of $294600\,\text{GHz}$) to the excitation frequency $\nu$ with respect to two atoms at $0\,\text{G}$ as described in section 5.1. The lines correspond to our numerical simulations, where the dashed line connects to a state which we cannot observe at $0\,\text{G}$ due to the selection rule $\Delta F=0,\pm1$. As one can see from Fig. 5.7, the rotational interaction which leads to the splitting of $6B_v$ between $J=0$ and $J=2$ dominates where the rotational energy is proportional to $B_v\cdot J(J+1)$. The rotational structure is dominant because the Zeeman interaction and the hyperfine interaction have diagonal elements proportional to $\Omega_I\cdot\Omega$. Hence, their contribution is off-diagonal and rather small. Because of the weak hyperfine structure, $J$ remains a good quantum number and the $0_g^-$ component can be well described by the basis vector

$$|S,\Sigma,(J,I)F,M_F\rangle$$

where $J$ results from the coupling of the electronic spin and the molecular rotation ($\mathbf{J}=\mathbf{N}+\mathbf{S}$). Looking at Table 5.1, we find good agreement between theory and

# 5. One-photon spectroscopy of the $(1)\,^3\Sigma_g^+$ potential

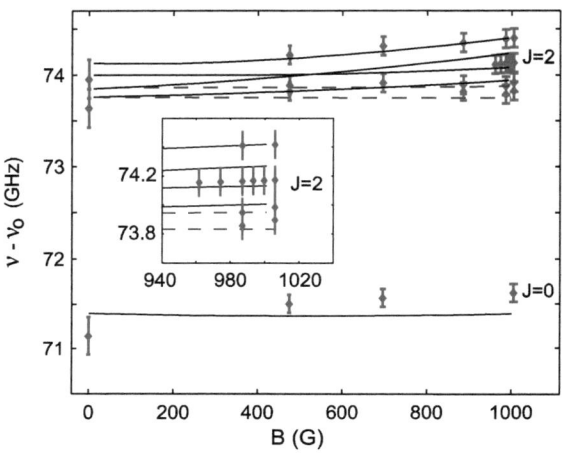

Figure 5.7: Excitation energy vs. magnetic field (in Gauss) for the $0_g^-$ state. The corresponding numerical values are given in table 5.1 at page 73. Energies are given with respect to two atoms at 0 G as discussed in section 6.1. Here we subtract an overall offset of $\nu_0 = 294600\,\mathrm{GHz}$. Experimental observations are plotted with corresponding uncertainties. Black lines are used for states that fulfill the selection rules at $B = 0$. The dashed line indicates a state that cannot be excited at $B = 0$ due to the selection rule for $F$. The experimental uncertainties are $\pm 60$MHz for measurements at $B > 0$, and $\pm 200$MHz at $B = 0$. Table 5.1 gives the quantum numbers at 0 G and 1005.8 G. $J = 1$ is not visible for parity reasons (page 65).

| ⟨J⟩ | ⟨I⟩ | ⟨F⟩ | B = 0 G | | ⟨I⟩ | ⟨F⟩ | B = 1005.8 G | |
|---|---|---|---|---|---|---|---|---|
| | | | $\nu_T - c$ (GHz) | $\nu_E - c$ (GHz) | | | $\nu_T - c$ (GHz) | $\nu_E - c$ (GHz) |
| 0.0 | 3 | 3.0 | 71.44 | 71.14 | 3.0 | 3.0 | 71.42 | 71.63 |
| 2.0 | 3 | 5.0 | 73.84 | n.o. | 3.0 | 4.2 | 73.83 | 73.83 |
| 2.0 | 1 | 3.0 | 73.84 | n.o. | 3.0 | 3.7 | 73.95 | 73.92 |
| 2.0 | 1 | 2.0 | 73.92 | n.o. | 1.0 | 2.9 | 74.00 | n.o. |
| 2.0 | 3 | 4.0 | 73.94 | n.o. | 3.0 | 3.3 | 74.13 | 74.13 |
| 2.0 | 3 | 3.0 | 74.06 | 73.64 | 1.0 | 2.1 | 74.27 | n.o. |
| 2.0 | 3 | 2.0 | 74.17 | 73.95 | 3.0 | 2.8 | 74.42 | 74.40 |

Table 5.1: We list here the calculated and measured excitation frequencies of the levels shown in Fig. 5.7 ($0_g^-(v' = 13)$) for the magnetic fields 0 G and 1005.8 G. The offset from the excitation frequency $\nu$ is $\nu_0 = 294600$ GHz. The subscripts $_T$ and $_E$ denote theoretical and experimental values, respectively. Along with the energies the expectation values for quantum numbers $J$, $I$ and $F$ are also given. All levels have $S = 1, \Sigma = 0$ and $M_F = 2$. Apparently, levels with $I = 1$ are not observed.

experiment if we assume that we cannot access states with $I = 1$. This seems reasonable because our Feshbach state has an $I = 1$ fraction of only 4%. It is possible to observe $J = 2$ for the following reason: Starting with the Feshbach state which has $N = 0$ we have the possibility to observe levels with $N = 1$ because of the selection rule $\Delta N \pm 1$. This state connects to $J = 2$ in a Hund's case c) coupling scheme.

Calculated and observed states together with their calculated quantum numbers $J$, $I$ and $F$ are also shown in table 5.1. Now we can understand the overall structure at 1005.8 G with the selection rules and the quantum numbers given in Table 5.1. For $J = 0$, we can only observe one state with $I = 3$ because we need a total angular momentum $F \geq 2$ at ≈294671 GHz. When we combine $J = 2$ with $I = 1$ ($I = 3$) we obtain states with $F = 2, 3$ ($F = 2, 3, 4, 5$). We should observe in total 6 states for $J = 2$. Our code shows that lines which belong to the same $F$ but differ in the $I$ quantum number are almost degenerate at 0 G. Increasing the magnetic field, lines cross and at 1000 G states with $I = 1$ and 2 are separated by ≈ 300 MHz. The states which cross always have different $I$ quantum numbers, and thus the interaction between these states has to be small. As can be read of from the expectation values in the table, for low magnetic fields the levels in the $0_g^-$ component are indeed well described by the quantum numbers $|S, \Sigma, (J, I)F, M_F\rangle$.

The contribution to the hyperfine interaction for states with $\Omega = 0$ is purely off-diagonal, allowing us, to determine the value of the effective parameter $(b_F - \frac{1}{3}c)$. Our theoreticians observed a weak dependence on this parameter and it turns out that its value may lie between $(200-1000)$ MHz, hence $(b_F - \frac{1}{3}c)$ is effectively undetermined from the available data. Higher experimental accuracy is required, to estimate the effective parameter

$(b_F - \frac{1}{3}c)$.

In Fig. 5.7 two of the calculated lines are dashed. They correspond to $F = 4, 5$ which are not observable at 0 G because of the selection rule $\Delta F = 0, \pm 1$. For larger magnetic fields, the Zeeman interaction mixes states with different $F$, and the two levels that correlate with $F = 4, 5$ at 0 G become observable. As the the magnetic field is increased, some lines cross. Because $I$ is a good quantum number, crossings between levels with different $I$ quantum numbers are not avoided.

In general, we see good agreement between experimental and fitted energy levels. Some discrepancies occur at zero magnetic field where the experimental data is less accurate.

### 5.4.2 $1_g$ spectrum and magnetic shifts

We now discuss the energy levels as a function of the magnetic field for the $1_g$-states. Calculated energy levels together with the experimental data for various magnetic fields are shown in Fig. 5.8. The y-axis corresponds (up to an offset of 294600 GHz) to the excitation energy with respect to two atoms at 0 G as described in setcion 5.1. The agreement between the observed and computed (theoretical) energies is rather good. The spin-spin splitting parameter $\lambda$ is well determined from the observed spectra because it only shifts the $1_g$ with respect to the $0_g^-$ spectrum as described in section 5.3. On the right hand side of the figure we also give the rotational quantum number $J$. For the $1_g$ state each rotational level is twofold degenerate because of the $+/-$ symmetry and we only have the restriction $J \geq \Sigma = 1$. It turns out that the maximum rotation which we can observe is $J = 4$. The Feshbach state has $N = 0$ which allows us to observe levels with $N = 1$ according to the selection rule $\Delta N \pm 1$. The strong second-order spin-orbit coupling in the excited state causes mixing of different $N$ states according to $\Delta N = 0, \pm 2$. It turns out that the excited $1_g$ state is a superposition of $N = 1$, $N = 3$ and $N = 5$. These states have a $J = 4$ component in a Hund's coupling case c) scheme as explained in chapter 4.

This strong second-order spin-orbit interaction in the excited state shows up due to coupling of the electronic spin momenta onto the molecular axis. Thus the vibrational level $v' = 13$ in the $(1)\,^3\Sigma_g^+$ potential is best described in a Hund's case c) coupling scheme with state vectors

$$|\Sigma, I, F, M_F\rangle.$$

$F$ results from the coupling between $J$ and $I$ to create a total angular momentum. $F$ and $M$ have the same meaning as in case e) except for the order of coupling. We also calculated the expectation values for $\Omega_I = |\Omega_{I_1} + \Omega_{I_1}|$. Except for the first state where we obtain $|\Omega_I| = 2.7$, it is not well defined. In other words, the basis vectors are close to Hund's case $c_\beta$) [Freed 1966]. At zero magnetic field and low rotation $J = 1, 2$ the quantum number $F$ for the total angular momentum is good, and we observe a large splitting between states which correspond to the same $F$ but different values of $I$ (Here

# 5. One-photon spectroscopy of the $(1)\,^3\Sigma_g^+$ potential

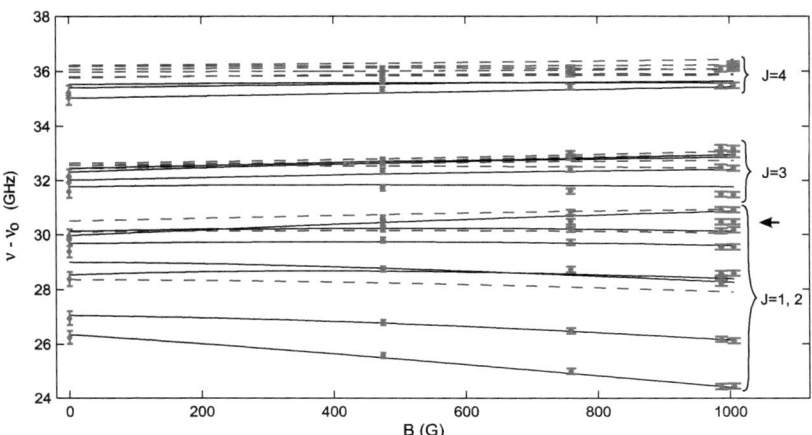

Figure 5.8: Zeeman diagram of the $1_g$ state ($v' = 13$). As in Fig. 6. an offset $\nu_0 = 294600\,\text{GHz}$ is subtracted from the excitation frequency $\nu$. The lines are calculations. Dashed lines indicate levels with $F > 3$ that can not be excited in our experiment at $B = 0$ due to the selection rule $F = 0, \pm 1$. Table 5.2 gives the corresponding quantum numbers at 0 G and 1005.8 G. The horizontal arrow is explained in the text.

$I = 1$ and $I = 3$.). This is in contrast to the $0_g^-$ states where these states were almost degenerate. Interestingly this splitting decreases with increasing $J$. For our last state with $J = 4$ we find a splitting of $\approx 100\,\text{MHz}$.

For increasing magnetic field, $F$ becomes worse. As in the $0_g^-$ state, we observe that only states with the same $I$ interact, which can be seen from various avoided crossings between such states. The quantum numbers $J$, $I$ and $F$ for the energy eigenstates at 0 G included in Fig. 5.8 are listed in Table 5.2.

For the lowest levels shown in Fig. 5.8, the hyperfine, the Zeeman, and the rotational interactions have comparable magnitudes. It is hard to recognize any regular pattern in the range from 24 GHz to about 37 GHz. Only at zero magnetic field and at low energies is the hyperfine interaction dominant, and hyperfine states with different $J$ do not overlap. As an example for the mixing of states with different $J$, one would expect that the line with lowest energy in the $v' = 13$ manifold has $J = 1$. Our simulations show that this is not the case, and we have an expectation value of $J = 1.4$.

The rotational energy starts to dominate over hyperfine and Zeeman energies for $J \geq 4$. This can be seen in the small splitting between the lines around 36 GHz (which have $J \cong 4$) and the few lines with energies less than 34 GHz (which have $J \cong 3$, see also table 5.2). In this regime, $J$ becomes a good quantum number. Indeed, looking at the quantum numbers in Table 5.2, we see the different hyperfine levels with $F = 2, 3, 4$ corresponding to $J = 1$ and $I = 3$. However, states with $J = 1, 2$ are strongly mixed and result in quantum numbers $J = 1.4$, $I = 3.0$ and $F = 2.2$ for the first state. The other two states which correspond to $F = 3, 4$ show considerable mixing with states from $J = 2$. Using this method, one can show that we would expect 10 states with $J = 1$ and 2 from theory. Furthermore, we have 8 states with $J = 3$, and 9 states with $J = 4$.

We still face some problems in the assignment of lines. Due to the close spacing of the energy levels at E $\simeq$ 36 GHz, it is impossible to establish a one-to-one correspondence between theoretical and calculated energies within this band. Furthermore, it seems that the line indicated by the horizontal arrow in Fig. 5.8 cannot be explained theoretically. In principle, it would be easiest to assign the lines at 0 G to reduce the number of lines because of the additional selection rule $\Delta F = 0, \pm 1$. Here, the problem is that fluctuations in the current cause losses when we sweep to 0 A. For this reason our signal to noise ratio is poor for the 0 G data and it is even more difficult to obtain spectra with high resolution on the order of a few MHz at 0 G.

| $\langle J \rangle$ | $\langle I \rangle$ | $\langle F \rangle$ | $\nu - c$ (GHz) |
|---|---|---|---|
| \multicolumn{4}{c}{$B = 0\,\text{G}$} |
| 1.2 | 3.0 | 2.0 | 26.34 |
| 1.4 | 3.0 | 3.0 | 27.06 |
| 2.7 | 3.0 | 4.0 | 28.36 |
| 1.2 | 1.0 | 2.0 | 28.53 |
| 2.0 | 3.0 | 2.0 | 29.00 |
| 2.0 | 3.0 | 3.0 | 29.68 |
| 1.9 | 1.0 | 2.0 | 29.98 |
| 2.1 | 3.0 | 5.0 | 30.12 |
| 2.3 | 1.0 | 3.0 | 30.13 |
| 2.0 | 3.0 | 4.0 | 30.51 |
| 3.0 | 3.0 | 2.0 | 31.77 |
| 3.0 | 3.0 | 3.0 | 32.02 |
| 3.0 | 1.0 | 2.0 | 32.31 |
| 2.9 | 3.0 | 4.0 | 32.44 |
| 3.0 | 1.0 | 3.0 | 32.45 |
| 3.1 | 3.0 | 6.0 | 32.56 |
| 3.2 | 1.0 | 4.0 | 32.57 |
| 3.0 | 3.0 | 5.0 | 32.64 |
| 3.9 | 3.0 | 2.0 | 35.04 |
| 3.9 | 3.0 | 3.0 | 35.41 |
| 4.0 | 1.0 | 3.0 | 35.54 |
| 3.9 | 3.0 | 4.0 | 35.78 |
| 4.0 | 1.0 | 4.0 | 35.81 |
| 4.0 | 1.0 | 5.0 | 35.99 |
| 3.9 | 3.0 | 5.0 | 36.07 |
| 4.1 | 3.0 | 7.0 | 36.19 |
| 4.0 | 3.0 | 6.0 | 36.22 |

Table 5.2: Calculated and measured excitation frequencies of the levels shown in Fig. 5.8 ($1_g(v' = 13)$) for the magnetic fields $0\,\text{G}$. The offset from the excitation frequency $\nu$ is $\nu_0 = 294600\,\text{GHz}$. Along with the energies, the expectation values for quantum numbers $J$, $I$ and $F$ are also given. All levels have $S = 1$, $\Sigma = 0$ and $M_F = 2$. Levels with $I = 1$ are probably not observable.

# 6 Precision spectroscopy of the $a\,^3\Sigma_u^+$ potential

Until recently, the $a\,^3\Sigma_u^+$ potential of the Rb$_2$ molecule has remained largely experimentally unexplored. Natural samples of Rb$_2$ are normally found in their $X\,^1\Sigma_g^+$ state from which the lowest triplet state is somewhat difficult to reach in an optical Raman process due to the selection rule $\Delta S = 0$ (Compare Eq. (4.24) and Eq. (4.26).). After we have analyzed the deeply bound states in the excited $(1)\,^3\Sigma_g^+$ potential, we are ready to perform precision spectroscopy of $^{87}$Rb$_2$ molecules in the $a\,^3\Sigma_u^+$ $(5S_{1/2} + 5S_{1/2})$ state, where we resolve almost all vibrational levels. This will be the topic of the current chapter, where I show how we map out the rotational, hyperfine, and Zeeman structure within the vibrational levels with an accuracy as high as 30 MHz. In our experiments, we use dark-state spectroscopy, where a gas of weakly bound Feshbach molecules is irradiated by two laser beams. Molecular losses, induced by one of the two lasers, are suppressed when the second laser is tuned into resonance with a bound state (see Fig. 6.1). Fitting a coupled-channel model to our experimental data, Prof. Eberhard Tiemann constructed an accurate Born-Oppenheimer $a\,^3\Sigma_u^+$ potential. This enables us to calculate the wave functions of triplet bound states as well as their binding energies to 60 MHz$\times h$ accuracy (where $h$ is Planck's constant) over the whole manifold of vibrational and low rotational (N<5) levels. The theory and data agree to the extent that further refinement of the theoretical model requires a reduction in the experimental uncertainty.

Molecular spectroscopy with cold atomic gases goes back to the beginnings of laser cooling [Weiner 1999, Jones 2006, Köhler 2006]. An experiment closely related to ours is the one by Araujo et al. [deAraujo 2003] where the Na$_2$ triplet ground state was explored using a magneto-optical trap combined with two-color photoassociation spectroscopy. Our spectroscopy also makes use of a two-photon transition, but starts from cold Feshbach molecules rather than from free atoms. Recent investigations of the $a\,^3\Sigma_u^+$ potential of Rb$_2$ include the work of [Lozeille 2006], [Beser 2009] and [Mudrich 2010]. Using one-color photoassociation of the $a\,^3\Sigma_u^+$ potential of laser-cooled $^{85}$Rb, [Lozeille 2006] put tight constraints on the position of the repulsive wall of the $a\,^3\Sigma_u^+$ potential but was not able to resolve the vibrational structure. The two other groups determined several ro-vibrational levels using fluorescence spectroscopy [Beser 2009] or pump-probe photoionization spectroscopy of Rb$_2$ formed on helium nanodroplets [Mudrich 2010]. Our work goes well beyond these measurements as we fully resolve hyperfine, rotational, and Zeeman structure for almost all vibrational states. Highly precise data of the asymptotic behavior of the coupled $a\,^3\Sigma_u^+$- $X\,^1\Sigma_g^+$ system is contained in the large set of observed Feshbach resonances [Marte 2002, Roberts 2001, Erhard 2003, Papp 2006] and in the two-photon photoassociation measurements of four weakly bound levels at zero magnetic field by [Wynar 2000]. Progress has also been made parallel to this work in the

# 6. Precision spectroscopy of the $a\,^3\Sigma_u^+$ potential

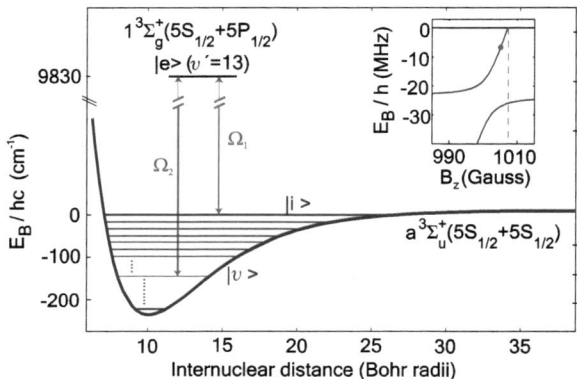

Figure 6.1: Dark-state spectroscopy scheme for the $^{87}$Rb$_2$ $a\,^3\Sigma_u^+$ potential. Lasers 1 and 2 couple the molecular levels $|i\rangle$ and $|v\rangle$ to the excited level $|e\rangle$ with Rabi frequencies $\Omega_{1,2}$, respectively. As excited level $|e\rangle$ we use a state in the vibrational level $v'=13$ which is best described by the state vector $|\Sigma, I, F, M_F\rangle = |1, 3, 2, 2\rangle$. The calculations for the excited state showed that the projection of the nuclear spin onto the internuclear axis is 2.7 which is close to the "ideal" value of 3. Laser 1 is kept on resonance while laser 2 can be tuned to any level of the $a\,^3\Sigma_u^+$ potential. Inset: Bound-state level of Feshbach molecules as a function of magnetic field $B_z$. The dashed line gives the position of the Feshbach resonance. The dot marks the Feshbach molecular state used in the experiments.

study of ultracold $Cs_2$ molecules [Danzl 2009, Mark 2009] and ultracold KRb molecules [Ni 2009, Ospelkaus 2010].

This chapter is organized as follows: Section 6.1 presents an overview of the setup for spectroscopy and typical dark-state spectroscopy scans. Section 6.2 summarizes the relevant quantum numbers and the assignment of the observed lines. Section 6.3 is a short summary of the coupled-channel model and the optimization procedure of the Born-Oppenheimer potentials. Section 6.4 discusses the progression of the substructure of the vibrational manifolds. I conclude the chapter with a summary of our experiments in section 6.5.

## 6.1 Experimental setup and dark state spectroscopy

The starting point for our experiments is a 50-$\mu$m-size pure ensemble of $3 \times 10^4$ weakly bound Feshbach molecules which have been produced from an atomic Bose-Einstein condensate of $^{87}$Rb by ramping over a Feshbach resonance at a magnetic field of 1007.4 G (1 G = $10^{-4}$ T) [Volz 2003]. This procedure is described in detail in chapter 3. We know from previous measurements that the Feshbach molecules are trapped in the lowest Bloch band of a cubic 3D optical lattice with no more than a single molecule per lattice site [Thalhammer 2006]. The lattice depth for the Feshbach molecules is $60\,E_r$, where $E_r = \pi^2\hbar^2/2ma^2$ is the recoil energy, with $m$ the mass of the molecules and $a = 415.22$ nm the lattice period. Such deep lattices prevent the molecules from colliding with each other, which suppresses collisional decay. We observe lifetimes of a few hundred ms. After producing the Feshbach molecules, the magnetic field is set to 1005.8 G where the spectroscopy is carried out. At this magnetic field, the binding energy of the Feshbach molecules is $4.4(3)$ MHz$\times h$ (see Fig. 6.1, inset).

Our dark state spectroscopy works as follows: The Feshbach molecules in state $|i\rangle$ are irradiated by simultaneous rectangular pulses from lasers 1 and 2. The pulses typically last 10 $\mu$s with Rabi frequencies $\Omega_1$ and $\Omega_2$, respectively (Fig. 6.1). We keep laser 1 resonant with the $|i\rangle - |e\rangle$ transition and at a power $I_1$ of about 0.1 mW ($\Omega_1 = 2\pi \times 0.3$ MHz) such that in the absence of laser 2 about half of the molecules are lost by spontaneous emission from $|e\rangle$. Laser 2, with its power $I_2$ (up to a few hundred mW), is scanned. As long as laser 2 does not hit an $|e\rangle - |v\rangle$ resonance, laser 1 will continue to induce losses. When an $|e\rangle - |v\rangle$ resonance occurs, the initial state $|i\rangle$ is projected onto a dark state $|\Psi_{dark}\rangle = (\Omega_2|i\rangle - \Omega_1|v\rangle)/\sqrt{\Omega_1^2 + \Omega_2^2}$. Molecules in this dark state are shielded from excitation to the short-lived level $|e\rangle$ [Lang 2009a]. This leads to a suppression of molecular losses. After the lasers are switched off, we measure the number of molecules via a reverse magnetic field sweep through the Feshbach resonance, dissociating the remaining molecules in $|i\rangle$ into atoms which are detected by absorption imaging. These measurements are destructive, and for each point in a scan, a fresh sample of Feshbach molecules has to be prepared.

As described in chapter 5, the level $|e\rangle$ has an excitation energy of about 295 THz$\times h$ with

6. Precision spectroscopy of the $a\,^3\Sigma_u^+$ potential

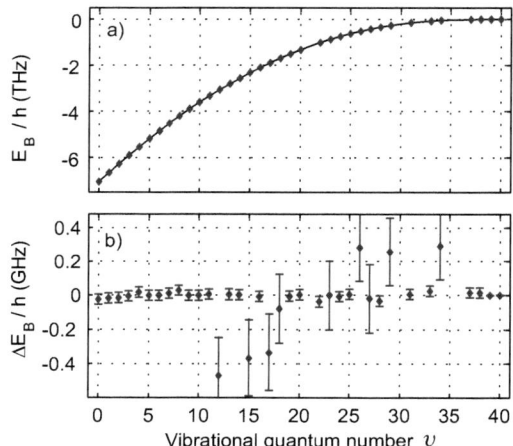

Figure 6.2: (a) Binding energies $E_B(v)$ for the state $a\,^3\Sigma_u^+$, where $v$ is the vibrational quantum number. The line is the result of a coupled-channel model calculation after optimization of the Born-Oppenheimer potential. Five levels were not measured. (b) Residues that is the difference between experimental data and theory. Large error bars belong to early measurements without simultaneous wave meter readings of both lasers.

respect to $|i\rangle$ and a natural width $\Gamma = 2\pi \times 8\,\text{MHz}$. Laser 1, a grating-stabilized diode laser, is Pound-Drever-Hall locked to a cavity, which is in turn locked to an atomic $^{87}$Rb-line. Laser 2, a Ti:sapphire laser, is free-running and typically drifts over a frequency range of a few MHz within seconds. Both lasers have a short-term laser linewidth of tens of kHz. The beams have a $1/e^2$ intensity waist radius of $130\,\mu\text{m}$ at the molecular sample, through which they propagate nearly collinearly. They are polarized parallel to the magnetic bias field $B_z$ (pointing in the vertical direction), and thus can only induce $\pi$ transitions. The frequencies of both lasers are automatically read out using a commercial wave meter (WS7 from HighFinesse). The binding energy $E_B$ of $|v\rangle$ minus the binding energy of our initial Feshbach state ($4.4\,\text{MHz} \times h$) corresponds to the frequency difference of the two lasers. The wave meter has a nominal accuracy of 60 MHz after calibration. Over the course of days we have observed drifts of $\pm 200$ MHz, for example by repeatedly addressing the same spectroscopic line. Over the length of a few experimental cycles (5 min), the wave meter is stable to within 10 MHz, which represents a random noise floor. Assuming a sufficiently smooth behavior of the wave meter, drifts typically affect the frequency measurements of laser 1 and laser 2 in a similar way, especially since the laser frequencies only differ by 3%. These common mode drifts cancel out in the binding energy to first approximation. Indeed, based on our experience with the wave meter where we have measured binding energies of a few sharp lines over an extended period of time and via various intermediate levels, we estimate the accuracy to reach 30 MHz. This includes peak position uncertainties due to variations in the number of molecules produced, as well as a frequency drift of laser 2 during the time between the laser pulse and wavelength measurement.

Fig. 6.2 a) shows the measured binding energies of the triplet potential as a function of the vibrational quantum number $v$ at a magnetic field of 1005.8 G. There are 41 vibrational states with binding energies ranging from $5\,\text{MHz} \times h$ to about $7038\,\text{GHz} \times h$. The vibrational splitting between the two lowest vibrational levels, $v = 0$ and $v = 1$, is about 393 GHz. Each vibrational state has hyperfine, rotational, and Zeeman substructure. This structure is spread out over a range of about 20 GHz as shown in Fig. 6.3 for the states $v = 0$ and 6. These spectra typically consist of roughly 1000 points corresponding to an average step size of 20 MHz. Each point represents one production and measurement cycle which takes 28 s.

In each spectrum of Fig. 6.3 we observe some 10 lines which vary markedly in linewidth. The width of each line is determined by the coupling between the levels $|e\rangle$ and $|v\rangle$, that is the Rabi frequency $\Omega_2$. This width of the dark resonance scales as $\Omega_2$ if laser 2 is fixed on resonance and laser 1 is scanned as explained in [Shore 1990]. Interestingly, for our measuring scheme this width scales as $\Omega_2^2$ [Lang 2009a] and not as $\Omega_2$, as one might expect. We have taken advantage of this enhanced broadening when searching for lines and vibrational manifolds, which otherwise can be like looking for a needle in a haystack. For example, for $v = 0$ and an intensity $I_2$ of a few hundred mW we reached linewidths of several GHz. The substructure of the desired vibrational level then appears essentially as a single broad line with a width of about 20 GHz. Once this level was found, the power was reduced in order to resolve its substructure.

# 6. Precision spectroscopy of the $a\,^3\Sigma_u^+$ potential

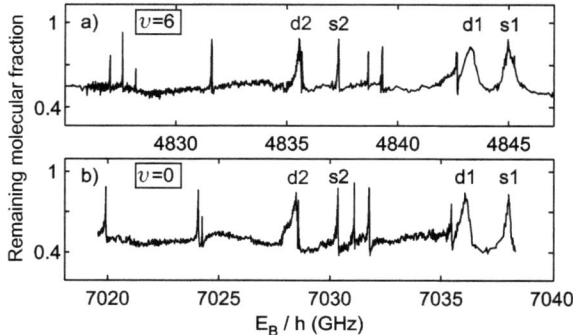

Figure 6.3: Scans of two vibrational levels within the $a\,^3\Sigma_u^+$ potential: a) $v = 6$, b) $v = 0$. Plotted is the remaining molecular fraction as a function of the binding energy $E_B$, which basically corresponds to the laser difference frequency. The scans were recorded using an excited state $|e\rangle$ with $1_g$ character. The labels s and d indicate rotational states $N = 0$ and 2, respectively. The numbers after the labels s or d indicate the position in the spectrum.

In general, we expect spectra of different vibrational levels to be very similar. Up to $v = 35$ this is indeed the case. For $v \geq 35$ the vibrational manifolds start to overlap, as the splitting between them becomes smaller than 20 GHz. Spectra a) ($v = 6$) and b) ($v = 0$) of Fig. 6.3 are clearly similar. Some visible differences are artifacts, e.g. it seems that in spectrum b) two narrow lines are missing at $E_B/h \approx 7021\,\text{GHz}$. This can be explained by the fact that the lines were narrower (width $\leq 10\,\text{MHz}$) than the local step size of that scan and thus escaped observation.

## 6.2 Quantum numbers and assignment

### 6.2.1 Quantum numbers

For a deeper understanding of the structure of the spectrum and its assignment we now discuss the relevant quantum numbers and selection rules.

For the weakly bound levels like the Feshbach state or the vibrational levels close to the atomic asymptote, Hund's coupling case e) with atomic quantum numbers is most appropriate. Here none of the angular momenta couple to the molecular axis. Using the nomenclature of chapter 4, the state vector is described by

$$|(f_a, f_b)f, l, F, M_F\rangle,$$

where $f_a, f_b$ are the total angular momenta for atoms $a$ and $b$, $f$ is the sum of both atomic angular momenta, $l$ is the mechanical rotation of the atomic pair, $F$ is the total angular momentum of the pair, and $M_F$ is its projection onto a space-fixed axis. At low magnetic fields the prepared Feshbach state $|i\rangle$ can be approximated by the state vector $|f_a = 2, f_b = 2, f = 2, l = 0, F = 2, M_F = 2\rangle$. At the magnetic field used in the experiment (1005.8 G), $F$ is no longer a good quantum number because of only weak coupling between $l$ and $f$. A more appropriate state vector is

$$|(f_a, f_b)f, m_f, l, m_l, M_F\rangle.$$

$M$ is a good quantum number, while for example, $f_a$ and $f_b$ have expectation values of about 1.79 instead of quantum numbers $f_a = f_b = 2$ for 0 G.

Due to the strong hyperfine coupling of Rb and the large exchange energy, deeply bound levels of the triplet state can be described by Hund's coupling case b$_\beta$) at low magnetic fields [Townes/Schawlow 1955], namely:

$$|N, (I, S)f, F, M_F\rangle,$$

where $I$ and $S$ are the total nuclear and electronic spin quantum numbers, $N$ is the molecular rotation including electron orbital angular momentum, and $F$ and $M_F$ have exactly the same meaning as in Hund's case e). The index $\beta$ indicates that the nuclear spin $I$ is not coupled to the molecular axis. Since both atoms are in the 5$S$ configuration of Rb, the molecular electronic orbital angular momentum is zero. This means that $N = l$ and that $f$ is the same as in Hund's case e). Here we get

$$|N, m_N, (I, S)f, m_f, M_F\rangle,$$

as an appropriate state vector for higher magnetic fields, where $M_F = m_N + m_f$. As discussed in section 4.7, one can show that owing to the antisymmetry of the molecular wave function with respect to nuclear exchange (nuclear spin of $^{87}$Rb $i = \frac{3}{2}$), molecules with even (odd) $N$ in the $a\,^3\Sigma_u^+$ state must have a total nuclear spin $I = 1$ or 3 ($I = 0$ or 2) [Townes/Schawlow 1955, Herzberg 1950]. For the $X\,^1\Sigma_g^+$ ground state this relation is reversed because it has $g$ symmetry in contrast to the $u$ symmetry of the triplet state. For large magnetic fields, $f$ loses its meaning as $S$ and $I$ start to decouple.

Expectation values for the total nuclear spin and the electron spin for our Feshbach level are $I = 1.56$ and $S = 0.76$, respectively. Thus, we have significant electronic singlet-triplet mixing and consequently also mixing of the basis vectors with different $I$. In contrast, the excited intermediate state $|e\rangle$ has well defined quantum numbers $S$ and $I$. Thus, $|e\rangle$ largely determines which quantum numbers the deeply bound $a\,^3\Sigma_u^+$ levels will have in the Raman transition.

The intermediate level $|e\rangle$ is located in the $v' = 13$ manifold of the $(1)\,^3\Sigma_g^+(5S_{1/2}+5P_{1/2})$ potential. Due to significant effective spin-spin interaction, the vibrational manifold is split into two components, $1_g$ and $0_g^-$, separated by 47 GHz. As intermediate level $|e\rangle$ we either choose a level with $1_g$ character or with $0_g^-$ character. Its rotational energy must

# 6. Precision spectroscopy of the $a\,^3\Sigma_u^+$ potential

be low because the Feshbach level has the lowest rotational quantum number $N = 0$. As rotation is low in $|e\rangle$, decoupling of $S$ and $I$ from the molecular axis by rotation is not yet important. We can then approximate the state $|e\rangle$ by quantum numbers for Hund's case $a_\alpha$) where the index $\alpha$ indicates that the nuclear spin $I$ is coupled to the molecular axis,

$$|S\Sigma, I\Omega_I, F, M_F\rangle\,.$$

The projections of the electronic and nuclear spin onto the molecular axis appear as quantum numbers $\Sigma$ and $\Omega_I$, respectively.

The $|e\rangle$ level with $1_g$ character is energetically the lowest within the $1_g(v = 13)$ manifold. It has the quantum numbers $S = 1$, $|\Sigma| = 1$, $I = 3$, $|\Omega_I| \approx 3$, $F \approx 2$ and $M = 2$ for low magnetic fields $B_z$. Its excitation energy is $294.6264(2)$ THz$\times h$ with respect to $|i\rangle$ at a field of $1005.8$ G.

The $|e\rangle$ level with $0_g^-$ character has an excitation energy of $294.6736(2)$ THz. It can be characterized by the quantum numbers $S = 1$, $\Sigma = 0$, $I = 3$, $M = 2$ and $F = 3$ at $B_z = 0$. Because of its low hyperfine coupling the better choice of basis vector here is

$$|S\Sigma, (JI)F, M_F\rangle\,,$$

where $J$ results from the coupling of the electronic spin and the molecular rotation ($\mathbf{J} = \mathbf{N} + \mathbf{S}$). For our $0_g^-$ $|e\rangle$ level $J$ is approximately zero. Our $1_g$ $|e\rangle$ state is a superposition of $N = 1$ and $3$.

As stated before, due to the laser polarization along the magnetic field only $\pi$ transitions are allowed, which results in the selection rule $\Delta M_F = 0$. Further, in a one-photon transition, parity has to change. The Feshbach state $|i\rangle$ has a total parity "plus" because it is a $\Sigma^+$ state and $(-1)^N = 1$ for $N = 0$. Thus, we can only address $|e\rangle$ levels with "minus" parity and $|v\rangle$ levels with "plus" parity. This means that the $|v\rangle$ level must have an even rotational quantum number $N = 0, 2, 4, \ldots$. For the $|e\rangle$ level, only quantum numbers $N = 1, 3, \ldots$ or superpositions of these are available. As the $(1)\,^3\Sigma_g^+$ state is well described in a Hund's case a) basis, $N$ is in general not a good quantum number for the $|e\rangle$ level. The selection rule $\Delta N = \pm 1$ for $N$ determines the range of reachable levels for $|v\rangle$ according to the superposition in $|e\rangle$. The selection rules $\Delta I = 0$ and $\Delta S = 0$ are important for the transition $|e\rangle$ to $|v\rangle$. For the transition $|i\rangle$ to $|e\rangle$ they are, however, nearly irrelevant since $I$ and $S$ are not good quantum numbers for $|i\rangle$.

## 6.2.2 Assignment of spectral lines

Figure 6.4 shows measured and calculated (based on the coupled-channel model - see section 6.3) lines of the $v = 6$ spectrum, where the excited state $|e\rangle$ with $0_g^-$ character was selected. The lines form three groups with the quantum numbers $f = 4, 3,$ and $2$, according to the hyperfine coupling of $S = 1$ and $I = 3$. This is a clear indication that the hyperfine energy is still dominant compared to the Zeeman energy.

# 6. Precision spectroscopy of the $a\,^3\Sigma_u^+$ potential

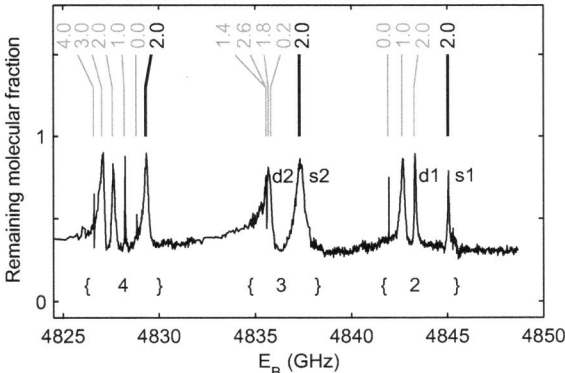

Figure 6.4: Scan of the hyperfine, rotational, and Zeeman structure in the $a\,^3\Sigma_u^+(v=6)$ manifold using an excited level $|e\rangle$ with $0_g^-$ character. The x-axis shows the binding energy $E_B$. The quantum number $f$ is shown below each of the three groups of lines. Thick black lines above the spectrum indicate states with $N=0$; grey lines correspond to states with $N=2$. The upper row of numbers indicates the expectation values of the magnetic quantum number $m_f$ at 1005.8 G. $m_f$ must be positive because of the restriction $m_f + m_N = m_F = 2$ and $-2 < m_N < 2$ for states with $N=2$.

Each group consists of one line with $N=0$ corresponding to a non-rotating molecule, and several lines with $N=2$ which are shifted to a lower binding energy by about 2 GHz×$h$ due to rotation. The fact that we do not observe lines with $N>2$ can be explained as follows. The excited state $|e\rangle$ has "minus" total parity. Since $J=0$ and $S=1$, $N$ must be equal to one and is also a good quantum number for this lowest level in the $0_g^-$ manifold. Thus when using the state with $0_g^-$ character only final states $|v\rangle$ with $N=0,2$ can be addressed in the $a\,^3\Sigma_u^+$ state due to the selection rule $\Delta N = \pm 1$.

The overall structure of each vibrational level can be understood with a relatively simple effective Hamiltonian [Dunn 1972]:

$$H = \frac{A}{2}\mathbf{S}\cdot\mathbf{I} + B_v\mathbf{N}^2 + \mu_B\,g_S\,S_z\,B_z$$

with the atomic hyperfine structure constant $A = 3.42$ GHz×$4\pi^2/h$ [Bize 1999], the total electronic and nuclear spin operator $\mathbf{S}$ and $\mathbf{I}$, respectively, and the operator for molecular rotation, $\mathbf{N}$. $B_v$ is the rotational constant of the desired vibrational level $v$. The last term describes the Zeeman effect of the electronic spin when exposed to an external magnetic field $B_z$ in the z-direction. The nuclear Zeeman term is neglected. The Zeeman effect can be evaluated for the case of strong hyperfine coupling such that $f$ (with $\mathbf{f} = \mathbf{S} + \mathbf{I}$) is still a good quantum number. Let us now consider a simple vector

## 6. Precision spectroscopy of the $a\,^3\Sigma_u^+$ potential

model, where the magnetic spin moment $\boldsymbol{\mu}$ moves in the magnetic field, which is caused by the electron spin and the nuclear spin. The moment $\boldsymbol{\mu}$ then precesses around the space-fixed direction of $\mathbf{f}$. Formally, we get the expectation value of the magnetic spin moment via[1]

$$\langle\boldsymbol{\mu}\rangle_f = \frac{\boldsymbol{\mu}\cdot\mathbf{f}}{|\mathbf{f}|^2}. \tag{6.1}$$

Using standard angular momentum techniques one can derive a Landé factor

$$g_f = \frac{g_S}{2}\left(1 + \frac{S(S+1) - I(I+1)}{f(f+1)}\right).$$

This Landé factor then leads to an effective Hamiltonian

$$H = \frac{A}{2}\mathbf{S}\cdot\mathbf{I} + B_v\mathbf{N}^2 + \mu_B\,g_f\,m_f\,B_z \tag{6.2}$$

For each $f$, the $N = 2$ group is shifted to lower binding energy by $6B_v$ compared to $N = 0$, corresponding to about 2 GHz for the $v = 6$ manifold (see Fig. 6.4). Each $N = 2$ group is split by the Zeeman energy according to the $m_f$ quantum number of each line. Only those lines which have quantum numbers where $M_F = 2 = m_f + m_N$ can be observed. The level $f = 3$ has a small $g_f$ factor (1/6 for $I = 3$), which gives rise to a small Zeeman splitting. Additionally, one sees that the splitting between $f = 4$ and $f = 2$, which would be about 12 GHz for the pure hyperfine part, is enlarged by a Zeeman contribution because of the different signs of $g_f$ for $f = 4$ and $f = 2$. In Fig. 6.4, the expectation values for $m_f$ resulting from the coupled channel model are given in the upper row and show that these lead to good quantum numbers for $f = 4$ and $f = 2$ but are less good for $f = 3$, where the hyperfine and electronic Zeeman energies are small. The mixing of states with different $m_f$ for $f = 3$ is not included in the simple model of Eq. 6.2. It could be included with an effective Hamiltonian of the form $\gamma\mathbf{N}\cdot\mathbf{S}$ (spin-rotation interaction).

As discussed before, for the intermediate state with $0_g^-$ character we only observed $N = 0$ and $N = 2$ levels in the $a\,^3\Sigma_u^+$ potential. This restriction does not necessarily apply when using the intermediate level $|e\rangle$ with $1_g$ character. The corresponding spectrum in Figure 6.3 a) shows the same $v = 6$ manifold as in Figure 6.4, only $|e\rangle$ has $1_g$ instead of $0_g^-$ character. It has additional lines (at $\sim$4838 GHz and at $\sim$4832 GHz) which match the predicted positions of $N = 4$ levels.

As mentioned in Section 6.2.1, the level $|e\rangle$ with $1_g$ character would be described by a superposition of $N = 1$ and $N = 3$ states in a Hund's case b) basis. We thus expect to see transitions to levels with $N = 0, 2, 4$ of the $a\,^3\Sigma_u^+$ state according to the selection rule $\Delta N = \pm 1$. In order to avoid confusion, we note that the $N = 0$ and $N = 2$ lines between 4825 GHz and 4830 GHz which are clearly visible in Figure 6.4 are weak or are

---

[1]This result is exact. If one uses tensor operators it corresponds to the reduced matrix element of $\mathbf{S}$ in the coupled scheme $|(SI)fm_f\rangle$.

not even observed in Figure 6.3. For the same reasons as discussed at the end of section 6.1, the experimental step size of about 10 MHz might have been too large compared to the narrow linewidths for these transitions to be seen.

It was not always necessary to carry out a complete scan as in Figure 6.4 in order to assign quantum numbers to observed lines in arbitrary vibrational levels. Often it was sufficient to measure a few characteristic lines and splittings and to compare them to the calculated spectrum. These data were then used to optimize the coupled channel model along with its $a\,^3\Sigma_u^+$ and $X\,^1\Sigma_g^+$ Born-Oppenheimer potentials.

## 6.3 Coupled-channel model and optimization of the $a\,^3\Sigma_u^+$ potential

In this section, I present the coupled-channel code [Krauss 1990, Dulieu 1995, Tiesinga 1998] which Prof. Eberhard Tiemann used for his theoretical investigations. The code can calculate all bound states of the $X\,^1\Sigma_g^+$ and $a\,^3\Sigma_u^+$ states, which correlate with the atomic asymptote $5^2S_{1/2}+5^2S_{1/2}$. The program has helped us with the search for lines as well as with their identification. Using our data, we are able to optimize the Born-Oppenheimer potential of the $a\,^3\Sigma_u^+$ state as well as to improve the potential of the $X\,^1\Sigma_g^+$ ground state given in [Seto 2000]. In the following, we briefly describe the model and explain how the $X\,^1\Sigma_g^+$ and $a\,^3\Sigma_u^+$ Born-Oppenheimer potentials are optimized. To cover the full range of experimental data by a single theoretical model, a coupled-channel analysis is highly adequate. It includes the calculation of the molecular bound states as well as the scattering resonances. It takes into account the $X\,^1\Sigma_g^+$ and $a\,^3\Sigma_u^+$ potential functions, the hyperfine coupling, the Zeeman interaction, rotation and the effective spin-spin interaction. Such a theoretical approach is described in several articles (for example [Pashov 2007]).

For the present analysis, we include our measurements covering 135 lines of the $a\,^3\Sigma_u^+$ state as well as data from other work. For the $X\,^1\Sigma_g^+$ ground state, we added data from Fourier transform spectroscopy by Seto et al. [Seto 2000] with more than 12000 lines. We also include measurements of Feshbach resonances for the three isotopologues $^{85}$Rb$_2$, $^{87}$Rb$_2$, and $^{85}$Rb$^{87}$Rb [Marte 2002, Roberts 2001, Erhard 2003, Papp 2006], the four asymptotic levels from [Wynar 2000], and measurements from Fourier transform spectroscopy for the $a\,^3\Sigma_u^+$ state reported by Beser et al. [Beser 2009].

The full Hamiltonian (cf. [Mies 2000, Laue 2002, Pashov 2007]) for a pair of atoms $A$ and $B$ can be written in the form

$$\begin{aligned}H = \; & T_n + U_X(R)P_X + U_a(R)(1-P_X) \\ & + a_a(R)\mathbf{s}_a \cdot \mathbf{i}_a + a_b(R)\mathbf{s}_b \cdot \mathbf{i}_b \\ & + \left[(g_{sa}s_{za} - g_{ia}i_{za}) + (g_{sb}s_{zb} - g_{ib}i_{zb})\right]\mu_B B_z \\ & + \frac{2}{3}\lambda(R)(3S_z^2 - S^2). \end{aligned} \qquad (6.3)$$

In the present case of the homonuclear molecule $^{87}$Rb$_2$ the parameters with index $a$ are equal to those of index $b$. The first term in the first line is the kinetic energy

## 6. Precision spectroscopy of the $a\,^3\Sigma_u^+$ potential

$T_n$ using the atomic masses from recent tables by G. Audi et al. [Audi 2003]. The next two terms describe the potential energies $U_X$ and $U_a$ for the motion of the atoms, where $P_X$ and $1 - P_X$ are projection operators on to the uncoupled states $X$ and $a$, respectively. $R$ is to the internuclear separation of the two atoms. The second line shows the hyperfine interaction between the atomic electron spins $\mathbf{s}_{a,b}$ and the atomic nuclear spins $\mathbf{i}_{a,b}$. The main contribution to the functions $a_{a,b}(R)$ is the Fermi contact term. The $R$ dependence of the hyperfine parameters accounts for several effects: It takes into account the electronic distortions of one atom by the other, that is the binding, and an effective coupling of the electron spin of one atom with the nuclear spin of the other atom. We start with $R$-independent atomic coupling constants taken from a compilation by Arimondo et al. [Arimondo 1977]. These constants are later refined by a simple Ansatz for the $R$ dependence which we discuss in Eq. (6.9). We neglect the nuclear quadrupole moment in the hyperfine interaction, which might come into play for deeply bound levels. The third line in Eq. (6.3) gives the Zeeman energy from the coupling of the electron spin and the nuclear spin with an external homogeneous magnetic field $B_z$ in the $z$ direction. The electronic and nuclear $g$ factors for the atomic ground state of the Rb isotopes are taken from the report in [Arimondo 1977]. This term couples states with different $f$ quantum numbers. The last line contains the spin-spin interaction which couples different $N$ states of basis b). It is formed by the total molecular spin $S$ and its projection on the molecule fixed axis $Z$. The parameter $\lambda$ is a function of $R$, one part of which has a $1/R^3$ dependence as a result of the magnetic dipole-dipole interaction. In addition, $\lambda$ contains contributions from second-order spin-orbit interactions [Tinkham 1954, Mies 1996]. The final analysis showed that such a contribution is significant within the achieved experimental accuracy. For example, this was important for the precise location of Feshbach resonances involving $l = 1$ and $l = 2$ levels.

The functional form of the two $X\,^1\Sigma_g^+$ and $a\,^3\Sigma_u^+$ Born-Oppenheimer potentials is split into three regions of the internuclear separation axis $R$: the short-range repulsive wall ($R < R_{\text{SR}}$), the asymptotic long-range region ($R > R_{\text{LR}}$), and the intermediate deeply bound region in-between. The analytic form of the potentials in the intermediate range, $U_{IR}$, is described by a finite power expansion of a nonlinear function $\xi$ which depends on the internuclear separation $R$,

$$U_{\text{IR}}(R) = \sum_{i=0}^{n} a_i\, \xi^i(R), \qquad (6.4)$$

$$\xi(R) = \frac{R - R_m}{R + b\, R_m}. \qquad (6.5)$$

Here the $a_i$ are fitting parameters (see Table 6.2). We choose $b$ and $R_m$ such that only a few parameters $a_i$ are needed for describing the steep slope on the short internuclear separation side and the much smaller slope on the large $R$ side. $R_m$ is chosen close to the value of the equilibrium separation. The potential is extrapolated for $R < R_{SR}$ by the short-range part $U_{SR}$ with

$$U_{SR}(R) = u_1 + u_2/R^{N_s}. \qquad (6.6)$$

# 6. Precision spectroscopy of the $a\,^3\Sigma_u^+$ potential

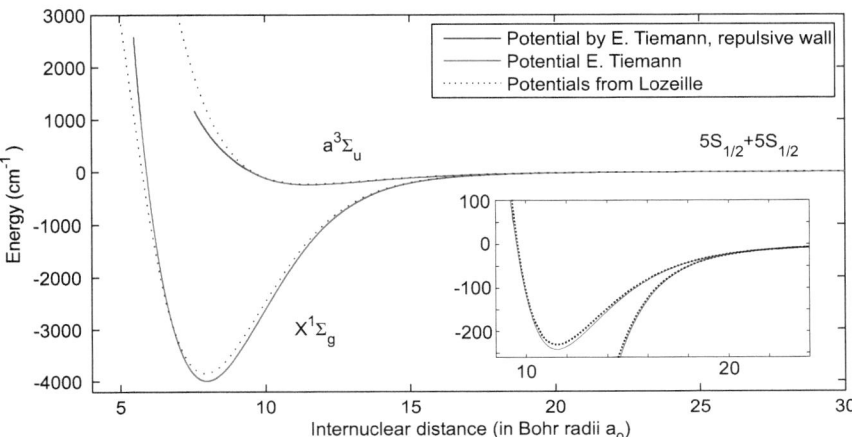

Figure 6.5: Optimized potentials for the singlet and triplet state. The dotted curves correspond to a calculation by [Lozeille 2006]. The solid red curves correspond to the intermediate and long-range part of the singlet and triplet potentials, whereas solid blue curves indicate the short-range part with the repulsive wall. It should be mentioned that our experiments are not sensitive to the actual shape of the repulsive wall, because we only measured bound states [Tiemann 2011]. The inset shows a zoom into the triplet potential.

We adjust the parameters $u_1$ and $u_2$ to get a continuous transition at $R_{\rm SR}$. The final fit uses $N_s \approx 4.5$ for both the $X\,^1\Sigma_g^+$ and $a\,^3\Sigma_u^+$ states.

For large internuclear distances ($R > R_{\rm LR}$), we adopted the standard long-range form of molecular potentials

$$U_{\rm LR}(R) = -C_6/R^6 - C_8/R^8 - C_{10}/R^{10} \pm E_{\rm exch}, \qquad (6.7)$$

where the exchange contribution given by [Smirnov 1965]

$$E_{\rm exch} = A_{\rm ex} R^\gamma \exp(-\beta R) \qquad (6.8)$$

is negative for the singlet and positive for the triplet potential. By adjusting the parameter $a_0$ in Eq. (6.4) we can assure a continuous transition from $U_{LR}$ to $U_{IR}$. As mentioned, the data on hand include the three different isotopologues, $^{85}$Rb$_2$, $^{87}$Rb$_2$, and $^{85}$Rb$^{87}$Rb. Using a model developed earlier for LiK [Tiemann 2009] we checked in the final calculations that the data are not sufficiently precise to extract deviations from the Born-Oppenheimer approximation. Thus we derive the potentials without any mass scaling correction. These potentials are shown in Fig. 6.5 (solid curves) together with

# 6. Precision spectroscopy of the $a\,^3\Sigma_u^+$ potential

an earlier calculation from Lozeille [Lozeille 2006]. The dissociation energy of the singlet potential changed by 200 cm$^{-1}$, the dissociation energy of the triplet potential changed by 10 cm$^{-1}$. Our experiments are not sensitive to the actual shape of the repulsive wall, because we only measured bound states. The inset shows a zoom into the triplet potential.

Having discussed the complete physical model, we are now ready to calculate all of the relevant bound-state energies and scattering properties to compare them with the experimental data. We decided to evaluate the Hamiltonian in Hund's basis e) as $f$ is still a relatively good quantum number. This is due to the fact that the hyperfine interaction is larger than the Zeeman interaction in our experiment. As the total electron spin is not a good quantum number in Hund's case e), the choice of this basis leads to significant non-diagonal matrix elements from the Born-Oppenheimer potentials, which are given for a pure singlet or triplet state.

We evaluate the free parameters of the model with a self-consistent iteration loop and alternate between: (i) coupled channel calculations of the full Hamiltonian and (ii) solving the Schrödinger equation separately for the states $X\,^1\Sigma_g^+$ and $a\,^3\Sigma_u^+$ using only the first line of Eq. (6.3) and applying the Numerov procedure. The coupled-channel calculations in step (i) are used to determine the hyperfine and Zeeman structure. From this, we construct hyperfine free spectroscopic data. This data is the input for the Born-Oppenheimer potential fits in step (ii). We incorporate the fitting routine for the plain Born-Oppenheimer potentials in step (ii). The optimization of the singlet and triplet potentials is done simultaneously as both potentials have a common asymptote. These asymptotic potentials are given in Eq. (6.7) with equal dispersion coefficients.

Normally, the iteration loop between the potential function fit and the coupled-channel calculation for producing hyperfine-structure-free data converges in a few steps, but we observed some systematic deviations between measurements and calculations. Specifically, the hyperfine splitting showed variations of a few percent within the vibrational ladder. Even though our experimental accuracy is not better than 30 MHz, such small variations are observable since the hyperfine splitting in $^{87}$Rb$_2$ of 12 GHz is so large.

In order to theoretically account for the hyperfine variations we extended the model. We changed the fixed atomic hyperfine parameters to a function of $R$. We chose a function which switches at a distance $R_0$ from the atomic value of the hyperfine constant to another value for a deeply bound dimer,

$$a_{a/b} = a_{Rb}\left(1 + \frac{c_f}{e^{(R-R_0)/\Delta R} + 1}\right). \tag{6.9}$$

Here $a_{Rb}$ is the atomic hyperfine constant, $c_f$ the fractional change of the constant and $R_0$ and $\Delta R$ describe the switching distance and its width, respectively. We also tried several other simple switching functions, which produced about equal fit quality. The function in Eq. (6.9) is easily applicable to other isotopes by introducing the proper atomic hyperfine constant, because the scaling parameter $c_f$ will be independent of the isotope. We chose $R_0 = 11.0\ a_0$ and $\Delta R = 0.5\ a_0$ (where $a_0 = 0.5292 \times 10^{-10}$ m is the Bohr radius), such that switching takes place approximately at the minimum of the $a\,^3\Sigma_u^+$ potential. From our fits, we obtained an amplitude $c_f = -0.0778$, which corresponds to a variation of the hyperfine coupling across the potential depth of up to 8%. Using $\langle R \mid qv \rangle = \Psi_{q,v}^{vib}(R)$, we can see that the influence of this hyperfine variation within an individual vibrational manifold will be smaller because of the averaging of Eq. (6.9) over the vibrational wave function

$$\left\langle q'v' \left| \frac{a_{a/b} - a_{Rb}}{a_{Rb}} \right| qv \right\rangle = \int \left(\Psi_{q',v'}^{vib}(R)\right)^* \frac{c_f}{e^{(R-R_0)/\Delta R} + 1} \Psi_{q,v}^{vib}(R)\ dR < c_f. \qquad (6.10)$$

As mentioned before, the spin-spin interaction also needs optimization in order to explain systematic shifts of Feshbach resonances in s-wave and p-wave scattering channels, as measured in [Marte 2002]. The spin-spin interaction couples different partial waves $l$ subject to the selection rule $l = 0, \pm 2$ such that resonances in a s-wave scattering channel involve bound states with $l = 0$ and $l = 2$. In higher order, this also involves states with $l = 4$ etc. The spin-spin interaction splits the resonances according to $|m_l|$, the projection quantum number of rotation on the space fixed axis. We use a simple functional form for $\lambda(R)$ in Eq. (6.3)

$$\lambda(R) = -\frac{3}{4}\alpha^2 \left(\frac{1}{R^3} + a_{SO} \exp\left(-b(R - R_{SO})\right)\right), \qquad (6.11)$$

that consists of two terms. The first term represents the magnetic dipole-dipole interaction and the second term is the second-order spin-orbit contribution. If Eq. (6.11) is given in atomic units, $\alpha$ is the fine structure constant. Since the few data at hand cannot be highly sensitive to the actual function, we adopted values for $b$ and $R_{SO}$ from a theoretical approach by Mies et al. for $Rb_2$ [Mies 1996] (b= 0.7196 $a_0^{-1}$ and $R_{SO}$= 7.5 $a_0$). We fitted the parameter $a_{SO}$, which yielded $a_{SO}$= -0.0416 $a_0^{-3}$. With these parameters, the second part in Eq. (6.11) contributes significantly to the effective spin-spin interaction in the internuclear separation interval $R < 20 a_0$. This affects bound levels of the triplet state that determine the Feshbach resonances. After optimization, we achieve an accuracy of 0.1 G. This correction to the constant $\lambda$ does not influence the description of the other bound states within their uncertainty.

Finally, we found that we could improve the fit by adding to the long-range formula Eq. (6.7) a term of the form $-C_{26}/R^{26}$ with an amplitude $C_{26}$, which contributes about a thousandth of the total long range energy at the connection point $R_{LR}$. The exponent of 26 was chosen so that the term is negligible outside a small region around the long range connection point $R = R_{LR}$.

The final parameter sets of the potentials are shown in Tables 6.1 and 6.2 for the singlet and triplet states, respectively. The derived potentials and the corrections defined previously agree very closely with all observations to within their experimental uncertainties, and the normalized standard deviation, that is standard deviation divided by experimental uncertainty, is close to one. In fact, depending on which of our experimental data sets (hyperfine-free spectra, binding energies from dark-state spectroscopy, or Feshbach resonances) we compare to the model calculations, the normalized standard deviations vary only slightly, ranging from 1.01 to 1.3. The normalized standard deviation from the joint calculation over the huge body of ro-vibrational energies (12459 data points from the states $X\,^1\Sigma_g^+$ and $a\,^3\Sigma_u^+$) is about 1.01, which is quite satisfactory.

As another result of our analysis, we were able to eliminate an ambiguity in the rotational assignment of the Fourier transform spectroscopy data reported by Beser *et al.* for the $a\,^3\Sigma_u^+$ state [Beser 2009]. The authors stated in the rotational assignment an ambiguity by $\Delta N = \pm 1$. We determined that the shift of the rotational quantum number must be $\Delta N = +1$. Afterwards we used the data from [Beser 2009] with that assignment for further fits. The results in Tables 6.1 and 6.2 include these data and will give better predictions of rotational levels with higher rotational quantum numbers.

## 6.4 Progression of vibrational levels and their substructure

In the following, we discuss interesting insights which we have gained from our analysis of the coupled system. In particular, we investigate the progression of several quantities in the vibrational ladder and the mixing of singlet and triplet states. We concentrate on the details of the $a\,^3\Sigma_u^+$ state because it was studied in full resolution of hyperfine and Zeeman energy over the whole vibrational ladder. Such a body of data does not exist for any other alkali-metal dimer.

### 6.4.1 Vibrational ladder and rotational progression

We return to the vibrational states shown in Fig. 6.2, which allows us to compare in detail the results of the optimized coupled-channel model with our experimental findings for all vibrational states.

The binding energies given in Fig. 6.2 (a) belong to the most deeply bound level in each vibrational manifold, that is, the state with quantum numbers $N = 0$, $f = 2$ (at 0 G) and $M_F = 2$. We will refer to this level as "s1". The "s1" level of $v = 0$ in the $a\,^3\Sigma_u^+$ potential is also the lowest bound state in that potential and has an observed binding energy of $(7038.067 \pm 0.050)\,\text{GHz} \times h$ at 1005.8 G with respect to the lowest atomic asymptote $f_a \approx 1, m_a = 1$ and $f_b \approx 1, m_b = 1$.

Fig. 6.2 (b) shows the residues between our experimental "s1" data and the optimized model. In general, the model agrees very well with the measurements to within the error bars. The data points with larger error bars belong to early measurements without simultaneous wave meter reading of both lasers, which leads to a significant increase in the experimental uncertainty (see section 6.1).

We now investigate the progression of the rotational splitting in the vibrational ladder. For this, we consider the "s1" level ($N = 0$) and its nearest neighbor in our spectra

| | $R < R_{\mathrm{SR}} = 3.126$ Å | |
|---|---|---|
| $N_s$ | | 4.53389 |
| $u_1^*$ | | $-0.638904880 \times 10^4$ cm$^{-1}$ |
| $u_2^*$ | | $0.112005361 \times 10^7$ cm$^{-1}$Å$^{N_s}$ |
| | $R_{\mathrm{SR}} \leq R \leq R_{\mathrm{LR}} = 11.000$ Å | |
| $b$ | | $-0.13$ |
| $R_m$ | | $4.209912760$ Å |
| $a_0$ | | $-3993.592873$ cm$^{-1}$ |
| $a_1$ | | $0.000000000000000000$ cm$^{-1}$ |
| $a_2$ | | $0.282069372972346137 \times 10^5$ cm$^{-1}$ |
| $a_3$ | | $0.560425000209256905 \times 10^4$ cm$^{-1}$ |
| $a_4$ | | $-0.423962138510562945 \times 10^5$ cm$^{-1}$ |
| $a_5$ | | $-0.598558066508841584 \times 10^5$ cm$^{-1}$ |
| $a_6$ | | $-0.162613532034769596 \times 10^5$ cm$^{-1}$ |
| $a_7$ | | $-0.405142102246254944 \times 10^5$ cm$^{-1}$ |
| $a_8$ | | $0.195237415352729586 \times 10^6$ cm$^{-1}$ |
| $a_9$ | | $0.413823663033582852 \times 10^6$ cm$^{-1}$ |
| $a_{10}$ | | $-0.425543284828921501 \times 10^7$ cm$^{-1}$ |
| $a_{11}$ | | $0.546674790157210198 \times 10^6$ cm$^{-1}$ |
| $a_{12}$ | | $0.663194778861331940 \times 10^8$ cm$^{-1}$ |
| $a_{13}$ | | $-0.558341849704095051 \times 10^8$ cm$^{-1}$ |
| $a_{14}$ | | $-0.573987344918535471 \times 10^9$ cm$^{-1}$ |
| $a_{15}$ | | $0.102010964189156187 \times 10^{10}$ cm$^{-1}$ |
| $a_{16}$ | | $0.300040150506311035 \times 10^{10}$ cm$^{-1}$ |
| $a_{17}$ | | $-0.893187252759830856 \times 10^{10}$ cm$^{-1}$ |
| $a_{18}$ | | $-0.736002541483347511 \times 10^{10}$ cm$^{-1}$ |
| $a_{19}$ | | $0.423130460980355225 \times 10^{11}$ cm$^{-1}$ |
| $a_{20}$ | | $-0.786351477693491840 \times 10^{10}$ cm$^{-1}$ |
| $a_{21}$ | | $-0.102470557344862152 \times 10^{12}$ cm$^{-1}$ |
| $a_{22}$ | | $0.895155811349267578 \times 10^{11}$ cm$^{-1}$ |
| $a_{23}$ | | $0.830355322355692902 \times 10^{11}$ cm$^{-1}$ |
| $a_{24}$ | | $-0.150102297761234375 \times 10^{12}$ cm$^{-1}$ |
| $a_{25}$ | | $0.586778574293387070 \times 10^{11}$ cm$^{-1}$ |
| | $R > R_{\mathrm{LR}}$ | |
| $C_6$ | | $0.2270032 \times 10^8$ cm$^{-1}$Å$^6$ |
| $C_8$ | | $0.7782886 \times 10^9$ cm$^{-1}$Å$^8$ |
| $C_{10}$ | | $0.2868869 \times 10^{11}$ cm$^{-1}$Å$^{10}$ |
| $C_{26}^{**}$ | | $0.2819810 \times 10^{26}$ cm$^{-1}$Å$^{26}$ |
| $A_{ex}$ | | $0.1317786 \times 10^5$ cm$^{-1}$Å$^{-\gamma}$ |
| $\gamma$ | | $5.317689$ |
| $\beta$ | | $2.093816$ Å$^{-1}$ |
| | Derived constants: | |
| equilibrium distance: | $R_e^X = 4.20991(5)$ Å | |
| electronic term energy: | $T_e^X = -3993.5928(30)$ cm$^{-1}$ | |

Table 6.1: Parameters of the analytic representation of the $X\,^1\Sigma_g^+$ state potential. The energy reference is the dissociation asymptote and the term energy $T_e^X$ is the depth of the potential. Parameters with $*$ are set for continuous extrapolation of the potential. See text for those with $**$.

## 6. Precision spectroscopy of the $a\,^3\Sigma_u^+$ potential

| | $R < R_{\mathrm{SR}} = 5.07$ Å |
|---|---:|
| $N_s$ | 4.5338950 |
| $u_1^*$ | $-0.619088543 \times 10^3$ cm$^{-1}$ |
| $u_2^*$ | $0.956231677 \times 10^6$ cm$^{-1}$Å$^{N_s}$ |
| | $R_{\mathrm{SR}} \leq R \leq R_{\mathrm{LR}} = 11.00$ Å |
| $b$ | $-0.33$ |
| $R_m$ | 6.0933451 Å |
| $a_0$ | $-241.503352$ cm$^{-1}$ |
| $a_1$ | $-0.672503402304666542$ cm$^{-1}$ |
| $a_2$ | $0.195494577140503543 \times 10^4$ cm$^{-1}$ |
| $a_3$ | $-0.141544168453406223 \times 10^4$ cm$^{-1}$ |
| $a_4$ | $-0.221166468149940465 \times 10^4$ cm$^{-1}$ |
| $a_5$ | $0.165443726445793004 \times 10^4$ cm$^{-1}$ |
| $a_6$ | $-0.596412188910614259 \times 10^4$ cm$^{-1}$ |
| $a_7$ | $0.654481694231538040 \times 10^4$ cm$^{-1}$ |
| $a_8$ | $0.261413416681972012 \times 10^5$ cm$^{-1}$ |
| $a_9$ | $-0.349701859112702878 \times 10^5$ cm$^{-1}$ |
| $a_{10}$ | $-0.328185277155018630 \times 10^5$ cm$^{-1}$ |
| $a_{11}$ | $0.790208849885562522 \times 10^5$ cm$^{-1}$ |
| $a_{12}$ | $-0.398783520249289213 \times 10^5$ cm$^{-1}$ |
| | $R > R_{\mathrm{LR}}$ |
| $C_6$ | $0.2270032 \times 10^8$ cm$^{-1}$Å$^6$ |
| $C_8$ | $0.7782886 \times 10^9$ cm$^{-1}$Å$^8$ |
| $C_{10}$ | $0.2868869 \times 10^{11}$ cm$^{-1}$Å$^{10}$ |
| $C_{26}^{**}$ | $0.2819810 \times 10^{26}$ cm$^{-1}$Å$^{26}$ |
| $A_{ex}$ | $0.1317786 \times 10^5$ cm$^{-1}$Å$^{-\gamma}$ |
| $\gamma$ | 5.317689 |
| $\beta$ | 2.093816 Å$^{-1}$ |
| | Derived constants: |
| equilibrium distance: | $R_e^a = 6.0940(10)$ Å |
| electronic term energy: | $T_e^a = -241.5034(30)$ cm$^{-1}$ |

Table 6.2: Parameters of the analytic representation of the $a\,^3\Sigma_u^+$ state potential. The energy reference is the dissociation asymptote and the term energy $T_e^a$ is the depth of the potential. Parameters with ∗ are set for continuous extrapolation of the potential. See text for those with ∗∗.

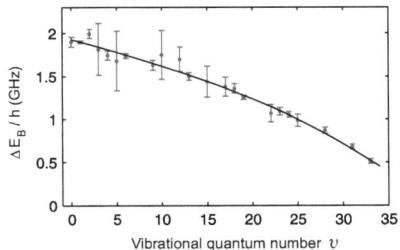

Figure 6.6: Energy splitting $\Delta E_B$ between the "s1" level ($N = 0$) and the "d1" level ($N = 2$) of the $a\,^3\Sigma_u^+$ state for a given vibrational quantum number $v$. Large error bars correspond to early measurements without simultaneous wave meter readings (see text). The continuous line is a calculation based on the coupled-channel model.

with ($N = 2$) which we call "d1" (see Figs. 6.3 and 6.4). The "s1" and "d1" levels both have $f = 2$ and $m_f = 2$. Thus, the splitting between them represents to a high degree rotational energy. The splitting decreases with increasing $v$, due to the fact that the mean distance between the Rb$_2$ nuclei (and hence the effective moment of inertia) increases as $v$ increases. We have directly observed this behavior in our experiments (see Fig. 6.6). There is very good agreement between the experimental data and the calculation (continuous line) using the optimized potential of the $a\,^3\Sigma_u^+$ state.

Besides the "s1" level ($N = 0, f = 2$) and the "d1" level ($N = 2, f = 2$) we also observe states with $N = 4$ and $f = 2$ for several low lying vibrational levels. The two lines in Fig. 3 (a) at 4839 GHz and the two lines in Fig. 3 (b) at 7032 GHz belong to $N = 4$. Observation of these levels improved the precision when fixing the position of the $a\,^3\Sigma_u^+$ potential minimum in terms of the internuclear distance, or the effective rotational constant $B_v$. This turned out to be important for the reassignment of the observations by Beser et al. [Beser 2009], as discussed at the end of section 6.3.

### 6.4.2 Hyperfine splitting and singlet-triplet mixing

It is instructive to investigate the progression of the multiplet structure of the vibrational levels. Figure 6.7 (a) shows calculated levels for $M_F = 2$ stacked on top of each other for increasing $v$ of the $a\,^3\Sigma_u^+$ potential. We concentrate on the triplet states ($S = 1$) with rotation $N = 0, 2$ and we restrict ourselves further to triplet levels with $I = 3$. Additionally, we show singlet levels ($S = 0$) with $I = 2$ that are located in the vicinity of the triplet levels. Thus, the typical stick spectrum of each vibrational manifold in the $a\,^3\Sigma_u^+$ state looks like the one in Fig. 6.4. There are three groups of lines with $f = 4, 3, 2$, respectively. The vibrational quantum number $v$ runs from 0 to 34, where the multiplets do not yet overlap for different vibrational levels.

# 6. Precision spectroscopy of the $a\,^3\Sigma_u^+$ potential

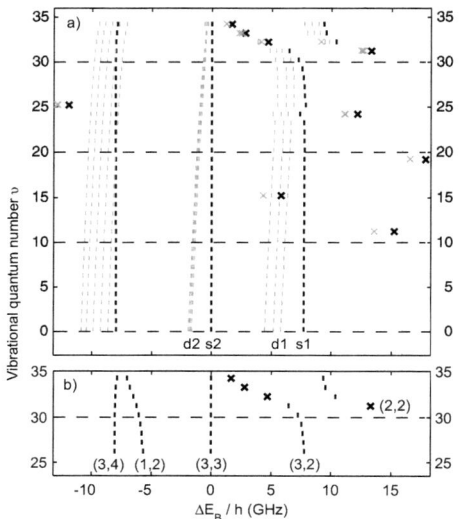

Figure 6.7: a) Progression of the substructure of the vibrational manifold at $B = 1005.8$ G. Shown are calculated $a\,^3\Sigma_u^+$ levels with quantum numbers $I \approx 3$, $M_F = 2$, $N = 0, 2$, and $v = 0...34$. Thick black lines correspond to $N = 0$, gray ones to $N = 2$. The "s2" level serves as energy reference ($\Delta E_B = 0$) in each vibrational substructure. In addition to the triplet levels, nearby singlet levels of the $X\,^1\Sigma_g^+$ potential are also shown (thick black crosses, $N = 0$; gray crosses, $N = 2$).

b) Symbols as in part a). To discuss the singlet-triplet mixing from a theoretical point of view, we now also include lines with $I \approx 1$ but only $N = 0$ for clarity. On the bottom, the approximate quantum numbers $(I, f)$ are given.

In order to properly stack the data, we have chosen the level with $N = 0$, $f \approx 3$ to be the energy reference ($\Delta E_B = 0$) for each vibrational level. We call this level "s2" (see also Figs. 3 and 4). This level is quite insensitive to mixing with singlet levels, which makes it a good reference because there is no singlet level with $f = 3$ and even parity for direct coupling. It is clear from Fig. 6.7 (a), that the multiplet structure of different vibrational manifolds is similar, at least from $v = 0$ to about $v = 30$, as expected from the simple model Hamiltonian in Eq. (6.2). The structure changes in a smooth and monotonic way with $v$. For each of the quantum numbers $f = 4, 3, 2$ the splitting between $N = 0$ and $N = 2$ decreases with increasing $v$ as discussed in section 6.4.1.

Appreciable mixing of a singlet and a triplet levels can occur when two levels with $f = 2$ are located close enough, in our case within a few GHz, corresponding to the strength of the hyperfine interaction. One effect of the mixing is a shift of the level positions due to level repulsion, clearly seen in Fig. 6.7. For $v = 24$, for example, the triplet lines of $f = 2$ are pushed to the left by about 0.4 GHz by the close singlet levels on the right-hand side. For $v < 20$ we also observe narrow coincidences between triplet and singlet levels, for example for $v = 15$. Here almost no perturbation appears in the graph, in contrast to the case $v = 24$. We attribute this to a significantly lower overlap of the vibrational wave functions of the singlet and triplet levels for the case $v = 15$ compared to that of $v = 24$. For vibrational quantum numbers $v > 30$, mixing is very strong and happens for every vibrational level. Because the long-range behavior of the $a\,^3\Sigma_u^+$ and $X\,^1\Sigma_g^+$ states is similar, the overlap of the wave functions will become large for high $v$. The vibrational spacing will become similar to the hyperfine splitting, and the state vectors here are best described by Hund's coupling case e), that is, quantum numbers of atom pairs.

Singlet-triplet mixing occurs not only for triplet molecules with $I = 3$ but also for $I = 1$. Figure 6.7 b) shows ($N = 0$) triplet levels with $I = 3$ and $I = 1$ for $v = 25$ to 34. The repulsion of the $I = 1$ levels from the singlet levels is clearly visible. The figure shows an avoided-crossing-like behavior for the levels on the right as a function of $v$, indicating strong mixing between $I = 3$ and $I = 2$. This means that the u/g symmetry is broken for these levels. Further, our calculations show that only levels with the same $f$ and $N$ quantum numbers mix considerably. The singlet lines shown here have $f = 2$ and $N = 0$. Apparently, despite the relatively strong magnetic fields of about 1000 G $f$ is still quite a good quantum number. Indeed, $f$ appears as quantum number in both state vectors for Hund's cases b) and e) and loses its meaning only for much higher magnetic fields. $N$ is good due to the small effective spin-spin interaction.

The shift in position due to mixing can be traced more clearly with a simple difference method discussed in the following. Fig. 6.8 shows the "discrete second derivative" $\delta^2 E_B(v)$ of the function for the binding energies $E_B(v)$, that is $\delta^2 E_B(v) = E_B(v+1) - 2E_B(v) + E_B(v-1)$. In other words, it is the difference between the neighboring energy splittings. The curve calculated from "s1" binding energies exhibits sudden jumps for particular vibrational levels. These are due to singlet-triplet mixing and the $v$ positions

## 6. Precision spectroscopy of the $a\,^3\Sigma_u^+$ potential

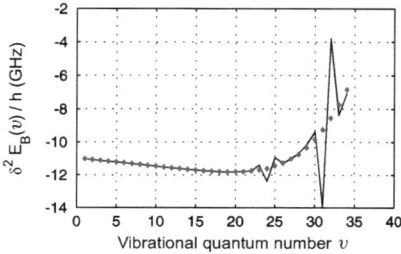

Figure 6.8: "Discrete second derivative" of $E_B(v)$ in the $a\,^3\Sigma_u^+$ potential, that is $\delta^2 E_B(v) = E_B(v+1) - 2E_B(v) + E_B(v-1)$. The solid curve connects the points from the calculated "s1" levels, while the diamonds correspond to the calculated "s2"-state.

are consistent with Fig. 6.7. In contrast, the $\delta^2 E_B(v)$ curve for the "s2" state is smooth, and thus does not indicate mixing with the singlet lines, which justifies its choice as energy reference in Fig. 6.7.

We have confirmed these singlet-triplet mixings experimentally. Fig. 6.9 shows scans of parts of the vibrational levels $v = 28, 31, 33$, where the intermediate level $|e\rangle$ with $1_g$ character was used (see also Fig. 6.3). The "s2" level is chosen to be the energy reference at $\Delta E_B = 0$ as before. For $v = 28$, we observe a structure similar to that of Figs. 6.3 and 6.4. Thus, we take the $v = 28$ spectrum as reference for the pure "triplet" case. In fact, from our calculations we see that the next singlet level is located about 30 GHz away. This detuning gives rise to only a very small mixing which slightly lowers the triplet character of the "s1"-level to 0.99. The singlet level obtains a triplet character of 0.02. However, due to the selection rule $\Delta S = 0$, the transition to the intermediate state $|e\rangle$ with its very pure triplet character would be simply too weak to detect.

This situation changes drastically for $v = 31$ and 33. Two additional lines are visible in the spectrum from singlet states with $M_F = 2$, $N = 0$ and 2. Here the singlet-triplet mixing is close to 40% which makes the singlet lines easily detectable. Additionally, for $v = 31$, the "s1" and "d1" components are shifted to lower values than expected from the reference spectrum $v = 28$ due to repulsion by the singlet component on the high energy side. In the case of $v = 33$ the singlet component pushes "s1" and "d1" in the opposite direction. Our calculations show that even the energy reference "s2" starts to show mixing, indicated by its reduced triplet character of 0.98. In parallel to the singlet-triplet mixing, the quantum number $I$ loses its meaning as well; for example, the "s1" level of $v = 33$ has an expectation value for $I$ of 2.75 instead of 3, due to a significant contribution of $I = 2$ from the singlet state. Comparing the stick-spectrum line positions with observed lines in Fig. 6.9, we note an excellent agreement between experiment and theory. We measured the singlet levels at the positions predicted from first calculations. This emphasizes the predictive power of the model.

# 6. Precision spectroscopy of the $a\,^3\Sigma_u^+$ potential

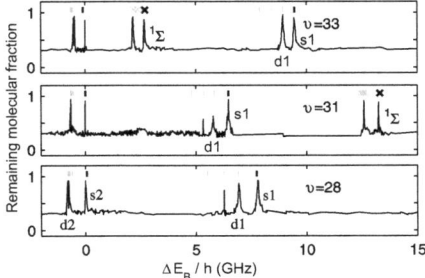

Figure 6.9: Observation of strongly perturbed singlet levels. We compare sections of line spectra for the vibrational levels $v = 28, 31$ and $33$. As in Fig. 6.7 the "s2"-level is chosen as energy reference $\Delta E_B = 0$. The lines (and crosses) above the experimentally observed spectrum are obtained from coupled-channel calculations. Black lines (crosses) represent $N = 0$ levels, while thin grey lines (crosses) correspond to $N = 2$ levels for $S = 1$ ($S = 0$ respectively). The lines that originate from the singlet states are labeled with "$^1\Sigma$".

The fact that we observe singlet-triplet mixing for relatively deeply bound levels with binding energies of a few hundred GHz$\times h$ is already quite interesting. In addition, it provides valuable information for fixing the energy position of the triplet levels with respect to the singlets with high precision. This is especially important for the large body of data of the singlet system (see section 6.3), which was obtained independent of the triplet state [Seto 2000].

### 6.4.3 Franck-Condon overlap $(1)\,^3\Sigma_g^+(v' = 13) \rightarrow a\,^3\Sigma_u^+(v)$

When scanning over all vibrational levels, the transition matrix elements for the transition from $|e\rangle$ to $|v\rangle$ are not constant but oscillate as a function of $v$. This oscillation is mainly due to variations in the Franck-Condon overlap between the $|e\rangle$, $v' = 13$ vibrational wave function and the vibrational wave functions $|v\rangle$ of the $a\,^3\Sigma_u^+$ potential. Figure 6.10 shows the normalized transition matrix element, $c_2 = \Omega_2(v)/(2\pi\sqrt{I_2})$, where $\Omega_2$ is the Rabi frequency and $I_2$ is the intensity of laser 2. $\Omega_2$ is determined from the measured width of the dark resonance using a three level model. Here, the Rabi frequency $\Omega_2$ is the only free parameter, and we fit the line shape to the dark resonance [Winkler 2007a]. For these measurements, we used the excited level $1_g$ as $|e\rangle$ and the "s1" levels as $|v\rangle$ of the $a\,^3\Sigma_u^+$ state. The transition matrix element $c_2$ varies from about 0.2 to 33 MHz/$\sqrt{\text{Wcm}^{-2}}$. In terms of a dipole moment, $\langle er \rangle = \Omega_2(v)/\sqrt{I_2} \times \hbar\sqrt{\epsilon_0 c/2}$, this corresponds to 0.05 to $8.0 \times 10^{-30}$ Cm.

In comparison, the amplitude for the transition between $|i\rangle$ and $|e\rangle$ is $\Omega_1/\sqrt{I_1} = 0.4$ MHz/

# 6. Precision spectroscopy of the $a\,^3\Sigma_u^+$ potential

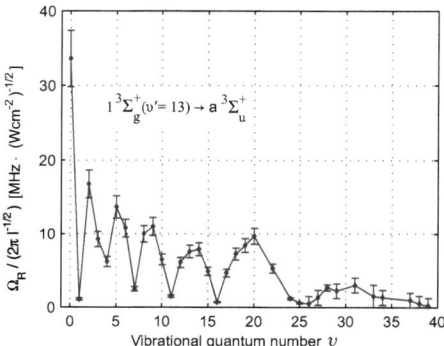

Figure 6.10: Normalized transition matrix element $\Omega_2/(2\pi\sqrt{I_2})$ between the excited level $|e\rangle$ ($v' = 13$, $1_g$ character) in the $(1)\,^3\Sigma_g^+$ potential and the "s1" level with vibrational quantum number $v$ in the lowest triplet potential $a\,^3\Sigma_u^+$.

$\sqrt{\mathrm{Wcm^{-2}}}$. We determine $\Omega_1$ from resonant excitation, measuring how quickly $|i\rangle$ molecules are lost for a given laser intensity $I_1$. The data in Fig. 6.10 can be used to fix the position of the $(1)\,^3\Sigma_g^+$ potential relative to the $a\,^3\Sigma_u^+$ potential by applying the Franck-Condon approximation.

## 6.5 Final analysis and reassignment of Feshbach resonances

The final result to be discussed in this section has developed from an interplay between experimental results and an optimized theory, which led in turn to a better understanding of the spectra or led to new predictions like the singlet-triplet mixture for deeply bound levels. By mainly optimizing the triplet Born-Oppenheimer potential, we obtain a model that can quite accurately predict all vibrational levels in the $a\,^3\Sigma_u^+$ and $X\,^1\Sigma_g^+$ potentials for different ranges of rotational quantum numbers. After reassigning recent data from Fourier transform spectroscopy [Beser 2009], Eberhard Tiemann extends the range of applicability to rotational states as high as $N = 70$ for the $a\,^3\Sigma_u^+$ potential. Because of this reassignment, molecular parameters like the equilibrium internuclear separation $R_e$ or the dissociation energy $D_e$ have changed significantly compared to values reported in [Beser 2009] (Reported values from Beser et al.: $R_e = 6.0690$ Å and $D_e = 241.4529$ cm$^{-1}$). The new values are given in Table 6.2. For rotational levels with $N \leq 4$ the model calculations for triplet levels of any vibrational level should have a precision similar to that of our measurements, that is, about 30 MHz. For higher $N$, this precision will increasingly degrade due to reduced accuracy of the data from [Beser 2009] of about 300 MHz for $N = 70$. Compared to [Seto 2000], the potential function of the $X\,^1\Sigma_g^+$ ground state is significantly improved in the region close to the atomic asymptote by

| isotope | $a_\text{singlet}$ | $a_\text{triplet}$ | $a_\text{lowest}$ | Rb + Rb $(f, m_f) + (f, m_f)$ |
|---|---|---|---|---|
| 87/87 | 90.35 | 99.04 | 100.36 | $(1,1) + (1,1)$ |
| 85/85 | 2720 | $-386.9$ | $-460.1$ | $(2,2) + (2,2)$ |
| 87/85 | 11.37 | 201.0 | 229.4 | $(1,1) + (2,2)$ |

Table 6.3: Scattering lengths (in units of Bohr radius, $a_0 = 0.5292 \times 10^{-10}$m). The scattering length of the energetically lowest hyperfine state is $a_\text{lowest}$.

including data on the mixed singlet-triplet levels of this study and data on Feshbach resonances from various other sources. The derived potential function can predict the deeply bound levels with an accuracy (about 50 MHz) comparable to that of the Fourier transform spectroscopy in [Seto 2000]. The asymptotic levels are accurate to a few MHz or better as their position is determined by the precisely measured location of Feshbach resonances. During the optimization process it became clear that we face systematic deviations in the description of the hyperfine structure. Using a hyperfine structure which is weakly dependent on the vibrational level, Eberhard Tiemann was able to resolve the problem. Furthermore, he realized that the modified hyperfine structure also influences the predictions of Feshbach resonances if one would like to reach accuracies on the order of 0.1 G.

Eberhard Tiemann improved the description of the large set of $^{87}$Rb Feshbach resonances with our final model as compared to Ref. [Marte 2002], where the original calculation was based only on a few selected resonances and a derived asymptotic form of the two potentials. In [Marte 2002], deviations as high as 2 G appeared and, for example, the resonances of asymptote $(f_a = 1, m_{f_a} = 0) + (f_b = 1, m_{f_b} = 1)$ were all calculated systematically too high. In the present model, these discrepancies disappear. The average deviation over all 46 resonances, given in [Marte 2002, Roberts 2001, Erhard 2003, Papp 2006], is about 0.15 G. In fact, using our optimized model, we were able to reassign three resonances reported by [Marte 2002]: 306.94 G to 2(0,1), 1137.97 G to 3(-1,1), and 729.43 G to 2(0,-1), where we use the abbreviated notation $l(m_{f_a}, m_{f_b})$. If one removes the $R$-dependence of the hyperfine coupling as given in Eq. (6.9), deviations of the order of 0.5 G appear. These vary according to the different $f$ levels which correlate to the Feshbach resonances. To improve the overall fit, one might be forced to include many-body effects in the collision process for detecting the resonances. The overall improvement of the long-range behavior also allows the whole set of scattering lengths for the three isotopologues to be calculated. Table 6.3 gives a selection of these calculations. The error of these data will be in the last digit shown. These values are consistent with earlier publications but more precise and, in general, internally consistent between the isotopologues.

# 7 Summary and Outlook

In this book, I present precision spectroscopy of ultracold $^{87}$Rb$_2$ molecules (See also [Strauss2011].). We perform one-photon spectroscopy of the first excited $(1)\,^3\Sigma_g^+$ state and Raman spectroscopy of the $a\,^3\Sigma_u^+$ ground state.

Driving bound-bound transitions between the Feshbach level and the first excited $(1)\,^3\Sigma_g^+$ state, we were able to resolve the low-lying vibrational ($v < 15$) and rotational structure. The accuracy of the position of the measured lines is about 60 MHz but the relative position of the lines with respect to each other is often much better and reaches a few tens of MHz. The dominating feature of the observed spectra is the splitting of the vibrational levels into $0_g^-$ and $1_g$ components which can be understood as an effect of strong second order spin-orbit coupling of the electrons. Furthermore, we investigate the hyperfine and Zeeman structure. In brief, the structure for the $0_g^-$ line is mainly determined by the vanishing spin component $\Omega = 0$, which leads to a very small hyperfine and Zeeman structure and a good quantum number $J$. In contrast, for the $1_g$ line hyperfine and Zeeman interactions are large for small rotations but are then averaged out at larger rotations, where the rotational splitting again determines the spectrum according to $J(J+1)$. The measurement of the Zeeman shifts for the excited state was technically challenging. In order to extract the magnetic shifts of the excited $(1)\,^3\Sigma_g^+$ state, we employed a technique where we kept track of the Feshbach molecule's quantum state. This technique enabled us to measure spectra at 0 G which fit the theory quite well. We compare our measurements with calculations from our collaborators Marius Lysebo and Leif Veseth in Oslo who use an effective Hamiltonian to calculate the level structure. That means that all numerical simulations are independent of the internuclear separation.

Despite the overall understanding of the level structure, we still observe systematic deviations between experiment and theory on the order of a few 100 MHz which should vanish in a more refined model. From our data we find that one of the two parameters for hyperfine interaction (the anisotropic part) could not be determined. If we want to measure the anisotropic hyperfine parameter in the future, the experimental data of the $0_g^-$ line will have to be measured with higher precision. It would be interesting to see whether the determined fit parameters for $\lambda$ (effective spin-spin interaction) and for the hyperfine contact interaction can be deduced from ab-initio calculations and the known atomic properties. Besides gaining a better insight into the level structure of the so far relatively unexplored $(1)\,^3\Sigma_g^+$ potential, the present work is helpful for future cold molecule experiments where the molecules need to be prepared in various well defined quantum states. We may then choose appropriate states in the $(1)\,^3\Sigma_g^+$ potential as intermediate states in an optical transfer scheme which starts from Feshbach molecules to maximize efficiency.

The $a\,^3\Sigma_u^+$ ground state is explored using high resolution dark-state spectroscopy. Finally, we obtain a Hamiltonian which we use to calculate all vibrational, rotational, and hyperfine levels in the $a\,^3\Sigma_u^+$ and $X\,^1\Sigma_g^+$ potentials. With our model, we can predict all deeply bound levels with rotational quantum number $N \leq 4$ in the $a\,^3\Sigma_u^+$ potential with an accuracy of about 30 MHz. For higher $N$, this precision will increasingly degrade due to reduced accuracy of the data from [Beser 2009] to about 300 MHz for $N = 70$. With the help of the singlet-triplet levels, it becomes possible to improve the $X\,^1\Sigma_g^+$ ground state model considerably. Using the data from [Seto 2000] and by including measurements on Feshbach resonances from various other sources, we can calculate the deeply bound levels with an accuracy of about 50 MHz. The asymptotic levels are accurate on the order of a few MHz or better as their position is determined by the precisely measured location of Feshbach resonances. Including the improved $R$-dependent hyperfine interaction, Eberhard Tiemann improved the description of the large set of $^{87}$Rb Feshbach resonances as compared to Ref. [Marte 2002]. In the present model, the average deviation over all 46 resonances, given in [Marte 2002, Roberts 2001, Erhard 2003, Papp 2006], is about 0.15 G.

Our model is valuable for our planned experiments on precision spectroscopy. The location of the molecular states is now well known in theory and experimental data will be mainly limited by the accuracy of the wave meter. Further experiments with higher precision could be also used to test the fundamental limits of the coupled-channel model, where we (1) assume the validity of the Born-Oppenheimer approximation, (2) use Zeeman terms with atomic parameters only, (3) a limited functional dependence in $R$ of the hyperfine and spin-spin interaction and (4) neglect quadrupole hyperfine coupling. Another set of experiments concerns the collisional behavior of ultracold triplet ground-state molecules. Up to now, it is not clear whether triplet molecules are stable when they collide with each other at temperatures below $1\mu$K. With this knowledge it might become possible to create a BEC of deeply bound triplet molecules. Jaksch et al. [Jaksch 2002] propose a method where they melt an optical lattice-induced Mott insulator. As we already created trapped molecules in the ro-vibrational triplet ground state [Lang 2008], we then have to lower the depth of the optical lattice to melt the Mott insulator and reach BEC. As we showed earlier, the molecules' dynamic is not trivial because it depends strongly on the lattice depth [Lang 2009].

A second path could use "classical" evaporation to throw the hottest molecules away. Here, we must drive transitions between the molecular triplet ground state and another molecular state which is not trapped magnetically. A detailed analysis is needed to identify the appropriate molecular states.

Triplet molecules have a magnetic moment, and thus a much richer level structure than singlet molecules. This structure is not only interesting by itself, but also serves as a potential starting point for exciting new experiments. Depending on the coupling between different triplet molecules, it could become possible to extend the scheme of Feshbach association towards dimer-dimer collisions to create weakly bound tetramers. In the future we also want to combine Feshbach association with a bichromatic superlattice. This lattice has a double-well structure where we can tune the potential depth and the interaction between two neighboring lattice-sites within one double well. Full control of the interactions between the two dimers hopefully allows management of the collisional behavior. This is an important step towards few-body physics, as already started in the experiments of [Zahzam 2006, Knoop 2009, Gross 2009, Pollack 2009, Knoop 2010, Ospelkaus 2010a] and [Ospelkaus 2010b]. Future experiments on tetramers will continue experiments on Efimov-states [Kraemer 2006, Ferlaino 2009, Zaccanti 2009] to further enter the field of ultracold chemistry [Staanum 2006, Krems 2005, Krems 2008].

Precision measurements of spectroscopic lines also open up the possibility to test for variation of the fundamental constants over time. For a review article see Chin et al. [Chin 2009]. An early proposal uses the scattering length to measure the proton-electron mass-ratio $m_p/m_e$ near narrow Feshbach resonances [Chin 2006]. Calculations based on single- and two-channel scattering models show that any variation of the mass ratio would propagate to the scattering length. They conclude that deviation should be observed when measuring nearby narrow magnetic or optical Feshbach resonances for $^6$Li or $^{133}$Cs. As there are also many Feshbach resonances in $^{87}$Rb [Marte 2002] which are known with high precision [Strauss 2010], it would be interesting to estimate the effect of a change in the proton-electron mass ratio for the case of $^{87}$Rb. Here, the control of the magnetic field strength and the homogeneity of the magnetic field are challenging but should be within the bounds of possibility [Chin 2009]. A similar article by Flambaum and co-workers [Flambaum 2007] suggests looking for variations of the fine structure constant and the proton-electron mass ratio $m_p/m_e$ using transitions between levels of different nature. For example, these levels can belong to different hyperfine structure components and transition frequencies typically lie in the microwave regime. Here, the required accuracy for $^{87}$Rb is about $10^{-5}$ Hz. As the ground state potentials of $^{87}$Rb$_2$ are $\Sigma$ states it is not possible to observe any drift of the fine structure constant because the energy (and thus, the drift) is proportional to $\Lambda \cdot \Sigma = 0$ to first order. A similar approach uses different electronic potentials and is given by DeMille [DeMille 2008]. In contrast to the proposal by Flambaum [Flambaum 2007], they are not sensitive to a change in the fine-structure constant $\alpha$ and only consider experiments to measure the proton-electron mass-ratio using a nearly degenerate pair of molecular vibrational levels. To detect a possible shift, a microwave field drives transitions between these fine-structure sublevels. Here, the sticking point is that the two levels belong to different electronic potentials. Furthermore they show that such a pair exists in Cs$_2$. As explained in chapter 6, $^{87}$Rb$_2$ exhibits several good singlet-triplet pair candidates. It would now be interesting to calculate the sensitivity of these levels in $^{87}$Rb$_2$ to estimate the effect of a change in the proton-electron mass ratio. Here, the measurement of the transition frequency with high precision is possible but not easy. All of this shows that molecules will have an exciting future!

# 8 Danksagung

An dieser Stelle möchte ich mich bei allen Personen bedanken, die mich bei meiner Doktorarbeit unterstützt haben. An erster Stelle steht hier natürlich Johannes Hecker Denschlag, der diese Arbeit hervorragend betreut hat und stets ein gutes Gespür dafür hatte, was die wichtigen physikalischen Fragestellungen sind. Er hatte stets Zeit für Diskussionen und hat wesentlich zum Gelingen dieser Arbeit beigetragen. Ebenso danke ich Rudi Grimm, der in Innsbruck eine herausragende Forschungsgruppe aufgebaut hat. Mit seinem Einsatz hat er es geschafft, eine Gruppe mit hervorragendem internationalem Ruf aufzubauen.

Ein riesiges Dankeschön geht natürlich auch an meine Vorgänger Florian, Klaus und Gregor, die das Rubidium-Projekt als Doktoranden aufgebaut und erweitert haben. Von ihnen habe ich sehr viel gelernt, und sie haben immer für ein tolles Arbeitsklima gesorgt, in dem es Spaß gemacht hat, zu forschen! Dank geht auch an die früheren Diplomanden Birgit, und Christian, die mit vollem Einsatz den Erfolg des Projektes vorangetrieben haben. Meinen Nachfolgern Björn und Markus danke ich für die tolle Zusammenarbeit in Ulm und wünsche Ihnen das nötige Quäntchen Glück und viele stabile Triplet Moleküle!

Im Besonderen möchte ich auch Tetsu danken, der mit seinem enormen Wissen maßgebend an dieser Arbeit beteiligt war! Sein Elan und sein Humor haben auch abseits der Physik für ein super Klima gesorgt. An dieser Stelle auch ein großes Dankeschön für das sorgfältige Korrekturlesen dieser Arbeit, und für die vielen Diskussionen über die englische Sprache.

Ein besonderer Dank geht auch an unseren "Theoretiker" Eberhard Tiemann, der mit seinem Wissen und seiner Freude am Erklären maßgeblich am Fortgang des Experiments beteiligt war. Seine Vorschläge, Singulett/Triplett Mischungen zu beobachten, und die Idee einer erneuten Auswertung der Feshbach Resonanzen haben mein Verständnis für Molekülphysik stark bereichert. Auch den Theoretikern Marius Lysebo und Leif Veseth danke ich für die tolle Zusammenarbeit und auch dafür, dass sie stets ein offenes Ohr für Fragen hatten.

Den "Barbies" Stefan und Arne, für die vielen schönen gemeinsamen Stunden in Innsbruck und Ulm, sowie all den anderen Gruppenmitgliedern in Innsbruck, die stets für ein angenehmes Arbeitsklima gesorgt haben: Vielen Dank für die vielen lustigen Stunden im Labor und auf den Bergen! Auch den neuen Kollegen in Ulm, die dafür gesorgt haben, dass das Einleben in Ulm sehr schnell von statten ging, ein großes Dankeschön.

# 8. Danksagung

Zu guter Letzt, möchte ich mich bei meinen Eltern bedanken, die mir diese Ausbildung ermöglicht haben, sowie meiner Freundin Birgit, die mich auch in den stressigen Zeiten immer unterstützt hat.

# Bibliography

[Adams 1997] C. S. Adams, and E. Riis, Laser cooling and trapping of neutral atoms, Prog. Quant. Electr. **21**, 1 (1997).

[Aldegunde 2008] J. Aldegunde, B. A. Rivington, P. S. Zuchowski, and J. M. Hutson, Hyperfine energy levels of alkali-metal dimers: Ground-state polar molecules in electric and magnetic fields, Phys. Rev. A **78**, 033434 (2008).

[Anderson 1995] M. H. Anderson, J. R. Ensher, M. R. Matthews, C. E. Wieman, and E. A. Cornell, Observation of Bose-Einstein condensation in dilute atomic vapor, Science **269**, 198 (1995).

[Andrè 2006] A. Andrè, D. DeMille, J. M. Doyle, M. D. Lukin, S. E. Maxwell, P. Rabl, R. J. Schoelkopf, P. Zoller, Nature Phys. **2**, 636-642 (2006).

[Andrews 1997] M. R. Andrews, C. G. Townsend, H.-J. Miesner, D. S. Durfee, D. M. Kurn, and W. Ketterle, Observation of Interference Between Two Bose Condensates, Science **275**, 637 (1997).

[Arimondo 1977] E. Arimondo, M. Inguscio, and P. Violini, Experimental determinations of the hyperfine structure in the alkali atoms, Rev. Mod. Phys. **49**, 31 (1977).

[Audi 2003] G. Audi, A. H. Wapstra and C. Thibault, The Image *NuBase* evaluation of nuclear and decay properties, Nuclear Physics A **729**, 337 (2003).

[Bai 2011] J. Bai, E. H. Ahmed, B. Beser, Y. Guan, S. Kotochigova, A. M. Lyyra, S. Ashman, C. M. Wolfe, J. Huennekens, Feng Xie, Dan Li, Li Li, M. Tamanis, R. Ferber, A. Droz- dova, E. Pazyuk, A. V. Stolyarov, J. G. Danzl, H.-C. Ngerl, N. Bouloufa, O. Dulieu, C. Amiot, H. Salami, and T. Bergeman, Phys. Rev. A **83**, 032514 (2011).

[Bahns 1996] J. T. Bahns, W. C. Stwalley, and P. L. Gould, Laser cooling of molecules: A sequential scheme for rotation, translation, and vibration, J. Chem. Phys. **104**, 9689 (1996).

[Bergeman 2011] Tom Bergeman, private communication.

[Bergmann 1998] K. Bergmann, H. Theuer, B. W. Shore, Coherent population transfer among quantum states of atoms and molecules, Rev. Mod. Phys. **70**, 1003 (1998).

[Beser 2009] B. Beser, V. B. Sovkov, J. Bai, E. H. Ahmed, C. C. Tsai, F. Xie, Li Li, V. S. Ivanov, and A. M. Lyyra, Experimental investigation of the $^{85}$Rb$_2$ $a^3\Sigma_u$ triplet

## Bibliography

ground state: Multiparameter Morse long range potential analysis, J. Chem. Phys. **131**, 094505 (2009).

[Bize 1999] S. Bize, Y. Sortais, M. S. Santos, C. Mandache, A. Clairon and C. Salomon, High-accuracy measurement of the $^{87}$Rb ground-state hyperfine splitting in an atomic fountain, Europhys. Lett. **45**, 558 (1999).

[Bjorklund 1983] G. C. Björklund, M. D. Levenson, W. Lenth, and C. Ortiz, Frequency Modulation Spectroscopy: Theory of Lineshapes and Signal-to-Noise Analysis, Appl. Phys. B **32**, 145 (1983).

[Black 2001] E. D. Black, An introduction to Pound Drever Hall laser frequency stabilization, Am. J. Phys. **69**, (1) (2001).

[Bourdel 2004] T. Bourdel, L. Khaykovich, J. Cubizolles, J. Zhang, F. Chevy, M. Teichmann, L. Tarruell, S. J. J. M. F. Kokkelmans, and C. Salomon, Experimental Study of the BEC-BCS Crossover Region in Lithium 6, Phys. Rev. Lett. **93**, 050401 (2004).

[Bradley 1995] C. C. Bradley, C. A. Sackett, J. J. Tollett, and R. G. Hulet, Evidence of Bose-Einstein Condensation in an Atomic Gas with Attractive Interactions, Phys. Rev. Lett. **75**, 1687 (1995).

[Brink/Satchler 1971] D. M. Brink, and G. R. Satchler, Angular Momentum, Clarendon Press Oxford 2nd. Ed. Reprinted (1971).

[Brown 1976] J. M. Brown and B. J. Howard, An approach to the anomalous commutation relations of rotational angular momenta in molecules, Molecular Physics **31**, 1517-1525 (1976).

[Brown/Carrington 2003] J. Brown, and A. Carrington, Rotational Spectroscopy of Diatomic Molecules, Cambridge Univerity Press (2003).

[Bunker 1968] P. R. Bunker, The Electronic Isotope Shift in Diatomic Molecules and the Partial Breakdown of the Born-Oppenheimer Approximation, J. Mol. Spectrosc. **28**, 422 (1968).

[Carr 2009] L. D. Carr, D. DeMille, R. V. Krems, and J. Ye, Cold and ultracold molecules: Science, technology and applications, New. J. Phys. **11**, 055049 (2009).

[Chin 2006] C. Chin, and V. V. Flambaum, Enhanced Sensitivity to Fundamental Constants In Ultracold Atomic and Molecular Systems near Feshbach Resonances, Phys. Rev. Lett. **96**, 230801 (2006).

[Chin 2009] C. Chin, V. V. Flambaum, and M. G. Kozlov, Ultracold molecules: new probes on the variation of fundamental constants, New J. Phys. **11**, 055048 (2009).

[Chin 2010] C. Chin, R. Grimm, P. Julienne, and E. Tiesinga, Feshbach resonances in ultracold gases, Rev. Mod. Phys. **82**, 1225 (2010).

[Chu 1985] S. Chu, L. Hollberg, J. E. Bjorkholm, A. Cable, and A. Ashkin, Three-dimensional viscous confinement and cooling of atoms by resonance radiation pressure, Phys. Rev. Lett. **55**, 48 (1985).

[Dalibard 1989] J. Dalibard, and C. C. Tannoudji, Laser cooling below the Doppler limit by polarization gradients: simple theoretical models, J. Opt. Soc. Am. B **6**, 11 (1989).

[Damski 2003] B. Damski, L. Santos, E. Tiemann, M. Lewenstein, S. Kotochigova, P. Julienne, and P. Zoller, Creation of a Dipolar Superfluid in Optical Lattices, Phys. Rev. Lett. **90**, 110401 (2003).

[Danzl 2008] J. G. Danzl, E. Haller, M. Gustavsson, M. J. Mark, R. Hart, N. Bouloufa, O. Dulieu, H. Ritsch, and H.-C. Nägerl, Quantum Gas of Deeply Bound Ground State Molecules, Science **321**, 1062 (2008).

[Danzl 2009] J. G. Danzl, M. J. Mark, E. Haller, M. Gustavsson, N. Bouloufa, O. Dulieu, H. Ritsch, R. Hart, and H.-C. Nägerl. Precision molecular spectroscopy for ground state transfer of molecular quantum gases. Faraday Discussions, **142**, 283 (2009).

[Danzl 2010] J. G. Danzl, Rovibronic Ground-State Molecules near Quantum Degeneracy. PhD thesis, Institut für Experimentalphysik, Universität Innsbruck (2010).

[Davis 1995] K. B. Davis, M. O. Mewes, M. R. Andrews, N. J. van Druten, D. S. Durfee, D. M. Kurn, and W. Ketterle, Bose-Einstein condensation in a gas of sodium atoms, Phys. Rev. Lett. **75**, 3969 (1995).

[deAraujo 2003] L. E. E. de Araujo, J. D. Weinstein, S. D. Gensemer, F. K. Fatemi, K. M. Jones, P. D. Lett, and E. Tiesinga, Two-color photoassociation spectroscopy of the lowest triplet potential of $Na_2$, J. Chem. Phys. **119**, 4 (2003).

[Deiglmayr 2008] J. Deiglmayr, A. Grochola, M. Repp, K. Mörtlbauer, C. Glück, J. Lange, O. Dulieu, R. Wester, and M. Weidemüller, Formation of Ultracold Polar Molecules in the Rovibrational Ground State, Phys. Rev Lett. **101**, 133004 (2008).

[DeMille 2002] D. DeMille, Quantum Computation with Trapped Polar Molecules, Phys. Rev. Lett. **88**, 067901 (2002).

[DeMille 2008] D. DeMille, S. Sainis, J. Sage, T. Bergeman, S. Kotochigova, and E. Tiesinga, Enhanced Sensitivity to Variation of $m_e/m_p$ in Molecular Spectra, Phys. Rev. Lett. **100**, 043202 (2008).

[Demtröder 2008] W. Demtröder, Laser spectrocopy, Springer, 4th Edition (2008).

[Doyle 2004] J. Doyle, B. Friedrich, R. V. Krems, F. Masnou-Seeuws, Enhanced Sensitivity to Variation of me/mp in Molecular Spectra, Eur. Phys. J. D **31**, 149 (2004).

[Drever 1983] Drever, R. W. P., J. L. Hall, F. V. Kowalski, J. Hough, G. M. Ford, A. Munley und H. Ward, Laser phase and frequency stabilisation using an optical resonator, Appl. Phys. B, **31** (1983).

# Bibliography

[Dürr 2005] S. Dürr, T. Volz, N. Syassen, G. Rempe, E. van Kempen, S. Kokkelmans, B. Verhaar, and H. Friedrich, Dissociation of Feshbach molecules into different partial waves Phys. Rev. A, **72** 052707 (2005).

[Dulieu 1995] O. Dulieu and P. S. Julienne, Coupled channel bound states calculations for alkali dimers using the Fourier grid method, J. Chem. Phys. **103**, 60 (1995).

[Dulieu 2009] O Dulieu, and C Gabbanini, The formation and interactions of cold and ultracold molecules: new challenges for interdisciplinary physics, Rep. Prog. Phys. **72**, 086401 (2009).

[Dunn 1972] T. M. Dunn, Nuclear Hyperfine Structure in the Electronic Spectra of Diatomic Molecules, in Molecular Spectroscopy: Modern Research, (Eds. K. N. Rao, C. W. Mathews), p. 231, Academic Press (1972).

[Edmonds 1960] A. R. Edmonds, Angular Momentum in Quantum Mechanics, Princeton University Press (1960).

[Erhard 2003] M. Erhard, H. Schmaljohann, J. Kronjäger, K. Bongs, and K. Sengstock, Bose-Einstein condensation at constant temperature, Phys. Rev. A **70**, 031602 (2003).

[Esslinger 1998] T. Esslinger, I. Bloch, and T. W. Hänsch, Bose-Einstein condensation in a quadrupole-Ioffe-configuration trap, Phys. Rev. A **58**, 2664 (1998).

[Fewell 1997] M. P. Fewell, B. W. Shore, and K. Bergmann, Coherent Population Transfer among Three States: Full Algebraic Solutions and the Relevance of Non Adiabatic Processes to the Transfer by Delayed Pulses, Austr. J. Phys. **50**, 281 (1997).

[Ferlaino 2009] F. Ferlaino, S. Knoop, M. Berninger, W. Harm, J. P. DÍncao, H.-C. Nägerl, and R. Grimm, Evidence for universal four-body states tied to an Efimov trimer, Phys. Rev. Lett. **102**, 140401 (2009).

[Flambaum 2007] V. Flambaum, and M. G. Kozlov, Enhanced sensitivity to the time variation of the fine-structure constant and $m_p/m_e$ in diatomic molecules, Phys. Rev. Lett. **99**, 150801 (2007).

[Foot 2005] C. J. Foot, Atomic Physics, Oxford University Press (2005).

[Freed 1966] K. F. Freed, On the Hyperfine Structure of InH and the Theory of the Hyperfine structure of Molecules in Hund's Case (C), J. Chem. Phys. **45**, 5 (1966).

[Frosch 1952] R. A. Frosch, and H. M. Foley, Magnetic Hyperfine Structure in Diatomic Molecules, Phys. Rev. **88**, 6 (1952).

[Greiner 2000] M. Greiner, Magnetischer Transfer von Atomen. Diplomarbeit, Max-Planck-Institut für Quantenoptik, Ludwig-Maximilians-Universität München (2000).

[Greiner 2001] M. Greiner, I. Bloch, T.W. Hänsch and T. Esslinger. Magnetic transport of trapped cold atoms over a large distance. Phys. Rev. A **63**, 031401 (2001).

[Greiner 2003] M. Greiner, C. A. Regal, D. S. Jin, Emergence of a molecular Bose-Einstein condensate from a Fermi gas, Nature **426**, 537 (2003).

[Greiner 2003a] M. Greiner, Ultracold quantum gases in three-dimensional optical potentials. Disseratation, Max-Planck-Institut für Quantenoptik, Ludwig-Maximilians-Universität München (2003).

[Grimm 2000] R. Grimm, M. Weidemüller, and Y. B. Ovchinnikov, Optical dipole traps for neutral atoms, Adv. At. Mol. OPt. Phys. **42**, 95 (2000).

[Gross 2009] N. Gross, Z. Shotan, S. Kokkelmans, and L. Khaykovich, Observation of universality in ultracold $^7$Li three-body recombination, Phys. Rev. Lett. **103**, 163202 (2009).

[Hänsch 1971] T. W. Hänsch, M. D. Levenson, and A. L. Schawlow, Complete Hyperfine Structure of a Molecular Iodine Line, Phys. Rev. Lett. **26**, 946 (1971).

[Hänsch 1975] T. W. Hänsch, and A. L. Schawlow, Cooling of gases by laser radiation, Opt. Commun. **13**, 68 (1971).

[Herbig 2003] J. Herbig, T. Kraemer, M. Mark, T. Weber, C. Chin, H.-C. Nägerl, R. Grimm, Preparation of a Pure Molecular Quantum Gas, Science **301**, 1510 (2003).

[Herzberg 1950] G. Herzberg, Molecular Spectra and molecular structure - Vol I, D. Van Nostrand company (1950).

[Hougen 1970] J. T. Hougen, The calculation of Rotational Energy levels and Rotational Line Intensities in Diatomic molecules, Nat. Bur. Stand., Monogr. 115, 52 pages (1930).

[Hund 1927] F. Hund, Zur Deutung der Molekelspektren. I., Z. Phys. **40**, 742 (1927).

[Hund 1927a] F. Hund, Zur Deutung der Molekelspektren. II., Z. Phys. **42**, 93 (1927).

[Hund 1933] F. Hund, Quantenmechanik des Atom- und Molekelbaues, in Handbuch der Physik **24**, 1, p. 628, Springer (1933).

[Hutson 2006] J. M. Hutson, P. Soldán, Molecule formation in ultracold atomic gases, Int. Rev. Phys. Chem. **25**, 497 (2006).

[Jaksch 2002] D. Jaksch, V. Venturi, J. I. Cirac, C. J. Williams, and P. Zoller, Creation of a molecular condensate by dynamically melting a Mott insulator, Phys. Rev. Lett. **89**, 040402 (2002).

[Jochim 2003] S. Jochim, M. Bartenstein, A. Altmeyer, G. Hendl, S. Riedl, C. Chin, J. Hecker Denschlag, and R. Grimm, Bose-Einstein Condensation of Molecules, Science **302**, 2101 (2003).

[Jones 2006] K. M. Jones, E. Tiesinga, P. D. Lett, and P. S. Julienne, Ultracold photoassociation spectroscopy: Long-range molecules and atomic scattering, Rev. Mod. Phys. **78**, 483 (2006).

[Kayama 1967] K. Kayama, and J. C. Baird, Spi-Orbit Effects and the Fine Structure in the $^3\Sigma_g^-$ Ground State of $O_2^*$, J. Chem. Phys. **46**, 2604 (1967).

[Ketterle 1999] W. Ketterle, D. S. Durfee, und D.M. Stamper-Kurn, Making, probing and understanding Bose-Einstein condensates, In: Bose-Einstein Condensation in Atomic Gases (Editors M. Inguscio, S. Stringari und C. E. Wieman) Vol. CXL in Proceedings of the International School of Physics Enrico Fermi. IOS Press (1999).

[Ketterle 1999a] W. Ketterle, and N. J. Van Druten, Evaporative cooling of trapped atoms, Adv. At. Mol. Opt. Phys **37**, 181 (1999).

[Knoop 2009] S. Knoop, F. Ferlaino, M. Mark, M. Berninger, H. Schöbel, H.-C. Nägerl, and R. Grimm, Observation of an Efimov-like trimer resonance in ultracold atom-dimer scattering, Nature Phys. **5**, 227 (2009).

[Knoop 2010] S. Knoop, F. Ferlaino, M. Berninger, M. Mark, H.-C. Nägerl, R. Grimm, J. P. DÍncao, and B. D. Esry, Magnetically controlled exchange process in an ultracold atom-dimer mixture, Phys. Rev. Lett. **104**, 053201 (2010).

[Köhler 2006] T. Köhler, K. Goral, and P. S. Julienne, Production of cold molecules via magnetically tunable Feshbach resonances, Rev. Mod. Phys. **78**, 1311 (2006).

[Kraemer 2006] T. Kraemer, M. Mark, P. Waldburger, J. G. Danzl, C. Chin, B. Engeser, A. D. Lange, K. Pilch, A. Jaakkola, H.-C. Nägerl, and R. Grimm, Evidence for Efimov quantum states in an ultracold gas of caesium atoms, Nature **440**, 315 (2006).

[Kramers 1929] H. A. Kramers, Zur Aufspaltung von Multiplett- S-Termen in zweiatomigen Molkülen. I, Zeitschrift fuer Physik **53**, 422 (1929).

[Kramers 1929a] H. A. Kramers, Zur Aufspaltung von Multiplett- S-Termen in zweiatomigen Molkülen. II, Zeitschrift fuer Physik **53**, 429 (1929).

[Krauss 1990] M. Krauss, and W. J. Stevens, Effective core potentials and accurate energy curves for $Cs_2$ and other alkali diatomics, J. Chem. Phys. **93**, 4236 (1990).

[Krems 2005] R. V. Krems, Molecules near absolute zero and external field control of atomic and molecular dynamics, Int. Rev. Phys. Chem. **24**, 99 (2005).

[Krems 2008] R. V. Krems, Cold controlled chemistry, Phys. Chem. Chem. Phys. **10**, 4079 (2008).

[Kristiansen 1986] P. Kristiansen and L. Veseth, Many-body calculations of hyperfine constants in diatomic molecules. I. The ground state of $^{16}OH$, J. Chem. Phys. **84**, 2711 (1986).

[Kronig 1930] R. de L. Kronig, Bandspectra and molecular structure, Cambridge University Press (1930).

[Lang 2008a] F. Lang, P.v.d. Straten, B. Brandstätter, G. Thalhammer, K. Winkler,

P.S. Julienne, R. Grimm, and J. Hecker Denschlag, Cruising through molecular bound-state manifolds with radiofrequency, Nature Phys. **4**, 223 (2008).

[Lang 2008] F. Lang, K. Winkler, C. Strauss, R. Grimm, and J. Hecker Denschlag, Ultracold Triplet Molecules in the Rovibrational Ground State, Phys. Rev Lett. **101**, 133005 (2008).

[Lang 2009] F. Lang, C. Strauss, K. Winkler, T. Takekoshi, R. Grimm, and J. Hecker Denschlag, Dark state experiments with ultracold, deeply-bound triplet molecules, Faraday Discussions **142**, 271-282 (2009).

[Lang 2009a] F. Lang, Coherent transfer of ultracold molecules: From weakly to deeply bound, PhD thesis, Institut für Experimentalphysik, Universität Innsbruck (2009).

[Laue 2002] T. Laue, E. Tiesinga, C. Samuelis, H. Knöckel, and E. Tiemann, Magnetic-field imaging of weakly bound levels of the ground-state $Na_2$ dimer, Phys. Rev. A **65**, 023412 (2002).

[Lefebvre-Brion/Field 2004] H. Lefebvre-Brion, and R. W. Field, The Spectra and Dynamics of Diatomic Molecules, Academic Press, Revised Edition (2002).

[Lett 1988] P. D. Lett, R. N. Watts, C. I. Westbrook, W. D. Phillips, P. L. Gould H. J. Metcalf, Observation of Atoms Laser Cooled below the Doppler Limit, Phys. Rev. Lett. **61**, 169 (1988).

[Lett 1993] P. D. Lett, K. Helmerson, W. D. Phillips, L. P. Ratliff, S. L. Rolston, and M. E. Wagshul, Spectroscopy of $Na_2$ by photoassociation of laser-cooled Na, Phys. Rev. Lett. **71**, 2200 (1993).

[Lindner 1984] A. Lindner, Drehimpulse in der Quantenmechanik, Teubner Studienbücher (1984).

[Lozeille 2006] J. Lozeille, A. Fioretti, C. Gabbanini, Y. Huang, H. K. Pechkis, D. Wang, P. L. Gould, E. E. Eyler, W. C. Stwalley, M. Aymar and O. Dulieu, Detection by two-photon ionization and magnetic trapping of cold $Rb_2$ triplet state molecules, Eur. Phys. J. D. **39**, 261 (2006).

[Lysebo 2009] M. Lysebo and L. Veseth, Ab initio calculation of Feshbach resonances in cold atomic collisions: s- and p-wave Feshbach resonances in $^6Li_2$, Phys. Rev. A **79** 062704 (2009).

[Lysebo 2010] M. Lysebo: Private communication.

[Marinescu 1995] M. Marinescu, and A. Dalgarno, Dispersion forces and lon-range electronic transition dipole moments of alkali-metal dimer excited states, Phys. Rev. A **52**, 52 (1995).

[Mark 2009] M. J. Mark, J. G. Danzl, E. Haller, M. Gustavsson, N. Bouloufa, O. Dulieu, H. Salami, T. Bergeman, H. Ritsch, R. Hart, and H.-C. Nägerl. Dark resonances for ground state transfer of molecular quantum gases. Applied Physics B, **95**, 219 (2009).

[Marquardt 1963] D. W. Marquardt, An Algorithm for Least-Squares Estimation of Nonlinear Parameters, Journal of the Society for Industrial and Applied Mathematics **11**, 431 (1963).

[Marte 2002] A. Marte, T. Volz, J. Schuster, S. Dürr, G. Rempe, E. G. M. van Kempen, and B. J. Verhaar, Feshbach Resonances in Rubidium 87: Precision Measurement and Analysis, Phys. Rev. Lett. **89**, 283202 (2002).

[Metcalf/van der Straten 1999] H. J. Metcalf, P. van der Straten, Laser Cooling and Trapping, Springer-Verlag, New York, (1999).

[Mies 1996] F. H. Mies, C. J. Williams, P. S. Julienne, and M. Kraus Estimating Bounds on collisional Relaxation Rates of Spin-Polarized $^{87}$Rb Atoms at Ultracold Temperatures, J. Res. NIST, **101**, 521 (1996).

[Mies 2000] F. H. Mies, E. Tiesinga, and P. S. Julienne, Manipulation of Feshbach resonances in ultracold atomic collisions using time-dependent magnetic fields, Phys. Rev. A **61**, 022721 (2000).

[Mudrich 2010] M. Mudrich, P. Heister, T. Hippler, C. Giese, O. Dulieu, and F. Stienkemeier, Spectroscopy of the triplet states of $Rb_2$ by femtosecond pump-probe photoionization of doped helium nanodroplets, Rev. Rev. A **80**, 042512 (2009).

[Mustelin 1963] N. Mustelin, On the coupling of angular momenta in diatomic molecules, with applications to the magnetic hyperfine structure. PhD thesis, Faculty of Mathematics and natural sciences of the Abo Akademi, Finland (1963).

[Ni 2008] K.-K. Ni, S. Ospelkaus, M. H. G. de Miranda, A. Pe'er, B. Neyenhuis, J. J. Zirbel, S. Kotochigova, P. S. Julienne, D. S. Jin, and J. Ye, A High Phase-Space-Density Gas of Polar Molecules, Science **322**, 5899 (2008).

[Ni 2009] K.-K. Ni, A Quantum Gas of Polar Molecules, PhD thesis, B.S. Creative Studies, University of California, Santa Barbara (2009).

[Ni 2010] K. K. Ni, S. Ospelkaus, D. Wang, G. Quemener, B. Neyenhius, M. H. G. de Miranda, J. L. Bohn, J. Ye, and D. S. Jin, Dipole collisions of polar molecules in the quantum regime, Nature **464**, 08953 (2010).

[Nikitin 1994] E. E. Nikitin, and R. N. Zare, Correlation diagrams for Hund's coupling cases in diatomic molecules with high rotational angular momentum, Mol. Phys. **82**, 85 (1994).

[Ospelkaus 2006] C. Ospelkaus, S. Ospelkaus, L. Humbert, P. Ernst, K. Sengstock, and K. Bongs, Ultracold heteronuclear molecules in a 3D optical lattice, Phys. Rev. Lett. **97**, 120402 (2006).

[Ospelkaus 2010] S. Ospelkaus, K. K. Ni, G. Quemener, B. Neyenhius, D. Wang, M. H. G. de Miranda, J. L. Bohn, J. Ye, and D. S. Jin, Controlling the Hyperfine State of the Rovibronic Ground-State Polar Molecules, Phys. Rev. Lett. **104**, 030402 (2010).

[Ospelkaus 2010a] S. Ospelkaus, K. K. Ni, D. Wang, M. H. G. de Miranda, B. Neyenhius, G. Quemener, P. S. Julienne, J. L. Bohn, D. S. Jin, and J. Ye, Quantum-State Controlled Chemical Reactions of Ultracold Potassium-Rubidium Molecules, Science **327**, (2010).

[Ospelkaus 2010b] S. Ospelkaus, K.-K. Ni, D. Wang, M. H. G. de Miranda, B. Neyenhuis, G. Quemener, P. S. Julienne, J. L. Bohn, D. S. Jin, and J. Ye, Quantum state controlled chemical reactions of ultracold potassium-rubidium molecules, Science **327**, 853 (2010).

[Papp 2006] S. B. Papp, and C. E. Wieman, Observation of Heteronuclear Feshbach Molecules from a $^{85}$Rb − $^{87}$Rb Gas, Phys. Rev. Lett. **97**, 180404 (2006).

[Pashov 2007] A. Pashov, O. Docenko, M. Tamanis, R. Ferber, H. Knöckel and E. Tiemann, Coupling of the $X^1\Sigma^+$ and $a^3\Sigma^+$ states of KRb, Phys. Rev. A **76**, 022511 (2007).

[Petrov 2004] D. S. Petrov, C. Salomon, and G. V. Shlyapnikov, Weakly Bound Dimers of Fermionic Atoms, Phys. Rev. Lett. **93**, 090404 (2004).

[Pollack 2009] S. E. Pollack, D. Dries, and R. G. Hulet, Universality in three- and four-body bound states of ultracold atoms, Science **326**, 1683 (2009).

[Press et al. 2007] William H. Press, S. A. Teukolsky, W. T. Vetterling and B. P. Flannery, Numerical Recipies, 3rd Ed. Cambridge University Press (2007).

[Pupillo 2008] G. Pupillo, A. Micheli, H.P. Büchler, P. Zoller, arXiv:0805.1896 (2008).

[Regal 2003] C. A. Regal, C. Ticknor, J. L. Bohn, and D. S. Jin, Creation of ultracold molecules from a Fermi gas of atoms, Nature **424**, 47 (2003).

[Roberts 2001] J. L. Roberts, J. P. Burke, N. R. Clausen, S. L. Cornish, E. A. Donley, and C. E. Wieman, Improved characterization of elastic scattering near a Feshbach resonance in $^{85}$Rb, Phys. Rev. A **64**, 024702 (2001).

[Rosa 2004] M. D. Rosa, Laser cooling molecule, Eur. Phys. J. D **31**, 395 (2004).

[Rose 1957] M. E. Rose, Elementary theory of angular momentum, John Whiley & Sons(1957).

[Sage 2005] J. M. Sage, S. Sainis, T. Bergeman, D. DeMille, Optical Production of Ultracold Polar Molecules, Phys. Rev. Lett. **94**, 203001 (2005).

[Sakurai 1995] J. J. Sakurai, Modern Quantum Mechanics, Addison-Wesley Publishing Company, Revised Edition (1995).

# Bibliography

[Schnell 2009] M. Schnell and G. Meijer, Cold Molecules: Preparation, applications, and challenges, Angew. Chem. Int. Ed. **48**, 6010 (2009).

[Schuenemann 1999] U. Schünemann, H. Engler, R. Grimm, M. Weidemüller, and M. Zielonkowskic, Simple scheme for tunable frequency offset locking of two lasers, Rev. Sci. Instr. **70**, 242 (1999).

[Seto 2000] J. Y. Seto, R. J. LeRoy, J. Vergès, and C. Amiot, Direct potential fit analysis of the $X^1\Sigma^+$ state of $Rb_2$: Nothing else will do!, J. Chem. Phys. **113**, 3067 (2000).

[Shore 1990] B. W. Shore, The Theory of Coherent Atomic Excitation, Wiley New York, (1990).

[Shuman 2010] E. S. Shuman, J. F. Barry, and D. DeMille, Laser cooling of a diatomic molecule, Nature , 09443 (2010).

[Smirnov 1965] B. M. Smirnov and M. S. Chibisov, Zh. Eksp. Teor. Fiz. 48, 939 (1965)[Sov. Phys. JETP 21,624 (1965)].

[Staanum 2006] P. Staanum, S. D. Kraft, J. Lange, R. Wester, and Matthias Weidemüller, Experimental Investigation of Ultracold Atom-Molecule Collisions, Phys. Rev. Lett. **96**, 023201 (2006).

[Steck 2003] D. A. Steck, Rubidium 87 D Line Data, http://steck.us/alkalidata, revised Version (2003).

[Stienkemeier 2006] F. Stienkemeier, and K. K. Lehmann, Spectroscopy and dynamics in helium nanodroplets, J. Phys. B: At. Mol. Opt. Phys. **39**, R127 (2006).

[Strauss 2010] C. Strauss, T. Takekoshi, F. Lang, K. Winkler, R. Grimm, E. Tiemann, and J. H. Hecker Denschlag, Hyperfine, rotational, and vibrational structure of the $a^3\Sigma_u^+$ state of $^{87}Rb_2$ , Phys Rev A **82**, 052514 (2010).

[Strauss2011] C. Strauss, Precision spectroscopy with ultracold $^{87}Rb_2$ triplet molecules, PhD thesis, Institut für Quantenmaterie, Universität Ulm, (2011).

[Takekoshi 1998] T. Takekoshi, B. M. Patterson, and R. J. Knize, Observation of optically trapped cold cesium molecules, Phys. Rev. Lett. **81**, 5105 (1998).

[Takekoshi 2011] T. Takekoshi, C. Strauss, F. Lang, J. Hecker Denschlag, M. Lysebo, and L. Veseth, Hyperfine, rotational and Zeeman structure of the lowest vibrational levels of the $^{87}Rb_2$ (1) $^3\Sigma_g^+$ state, Phys. Rev. A **83**, 062504 (2011).

[Thalhammer 2006] G. Thalhammer, K. Winkler, F. Lang, S. Schmid, R. Grimm, and J. Hecker Denschlag, Long-Lived Feshbach Molecules in a Three-Dimensional Optical Lattice, Phys. Rev. Lett. **96**, 050402 (2006).

[Thalhammer 2007] G. Thalhammer, Ultrakalte gepaarte Atome in kohärenten Lichtfeldern. PhD thesis, Institut für Experimentalphysik, Universität Innsbruck, (2007).

[Theis 2005] M. Theis, Optical Feshbach Resonances in a Bose-Einstein Condensate, PhD thesis, Institut für Experimentalphysik, Universität Innsbruck, (2005).

[Tiemann 2009] E. Tiemann, H. Knöckel, P. Kowalczyk, W. Jastrzebski, A. Pashov, H. Salami and A. J. Ross, Coupled system $a^3\Sigma^+$ and $X^1\Sigma^+$ of KLi: Feshbach resonances and corrections to the Born-Oppenheimer approximation, Phys. Rev. A **79** 042716 (2009).

[Tiemann 2011] E. Tiemann, Private communication.

[Tiesinga 1998] E. Tiesinga, C. J. Williams, and P. S. Julienne, Photoassociative spectroscopy of highly excited vibrational levels of alkali-metal dimers: Green-function approach for eigenvalue solvers, Phys. Rev. A **57**, 4257 (1998).

[Tinkham 1954] M. Tinkham, and M. W. P. Strandberg, Theory of the Fine Structure of the Molecular Oxygen Ground State, Phys. Rev. **97**, 937 (1955).

[Toennies 1998] J. P. Toennies and A. F. Vilesov, Spectroscopy Of Atoms And Molecules in Liquid Helium, Annu. Rev. Phys. Chem. **49**, 1 (1998).

[Townes/Schawlow 1955] C. H. Townes, and A. L. Schawlow, Microwave Spectroscopy, McGraw-Hill Book Company (1955).

[Van de Meerakker 2008] S. van de Meerakker, T. Bethlem, and G. Meijer, Taming molecular beams, Nature Physics **4**, 595 (2008).

[VanVleck 1951] J. H. Van Vleck, The Coupling of Angular Momentum Vectors in Molecules, Rev. Mod. Phys. **23**, 213 (1951).

[Veseth 1976] L. Veseth, The Hyperfine Structure of Diatomic Molecules: Hund's Case ($c_\alpha$), J. Mol. Spectroc. **59**, 51 (1976).

[Veseth 1976a] L. Veseth, Theory of High-Precision Zeeman Effect in Diatomic Molecules, J. Mol. Spectrosc. **63**, 180 (1976).

[Viteau 2008] M. Viteau, A. Chotia, M. Allegrini, N. Bouloufa, O. Dulieu, D. Comparat, and P. Pillet, Optical Pumping and Vibrational Cooling of Molecules, Science **321**, 232 (2008).

[Volz 2003] T. Volz, S. Dürr, S. Ernst, A. Marte, and G. Rempe, Characterization of elastic scattering near a Feshbach resonance in $^{87}$Rb, Phys. Rev. A **68** 010702(R) (2003).

[Volz 2006] T. Volz, N. Syassen, D. M. Bauer, E. Hansis, S. Dürr, and G. Rempe, Preparation of a quantum state with one molecule at each site of an optical lattice, Nature Physics **2**, 692 (2006).

[Walls/Milburn 1994] D. F. Walls and G. J. Milburn, Quantum Optics, Springer- Verlag, Berlin, (1994).

[Wang 1997] H. Wang, P. L. Gould, and W. C. Stwalley, Long-range interaction of the $^{39}$K(4s)+$^{39}$K(4p) asymptote by photoassociative spectroscopy. I. The $0_g^-$ pure longe-range state and the long-range potential constants, J. Chem. Phys. **106**, 19 (1997).

[Weiner 1999] J. Weiner, V. S. Bagnato, S. Zilio and P. S. Julienne, Experiments and theory in cold and ultracold collisions, Rev. Mod. Phys. **71**, 1 (1999).

[Wilson 1955] E. B. Wilson, J. C. Decius, P. C. Cross, Molecular Vibrations The Theory of Infrared and Raman Vibrational Spectra, McGraw-Hill Book Company, (1955).

[Wineland 1975] D. J. Wineland and H. Dehmelt, Proposed $10^{14}$ $\Delta\nu < \nu$ laser fluorescence spectroscopy on Tl$^+$ mono-ion oscillator III, Bull. Am. Phys. Soc. **20**, 637 (1975).

[Winkler 2002] K. Winkler, Ultracold Molecules and Atom Pairs in Optical Lattice Potentials. Diplomarbeit, Institut für Experimentalphysik, Universität Innsbruck (2002).

[Winkler 2006] K. Winkler, G. Thalhammer, F. Lang, R. Grimm, J. Hecker Denschlag, A. J. Daley, A. Kantian, H. P. Büchler, and P. Zoller, Repulsively bound atom pairs in an optical lattice, Nature **441**, 853 (2006).

[Winkler 2007] K. Winkler, Molecules and Atom Pairs in Optical Lattice Potentials. PhD thesis, Institut für Experimentalphysik, Universität Innsbruck (2007).

[Winkler 2007a] K. Winkler, F. Lang, G. Thalhammer, P. v. d. Straten, R. Grimm, and J. Hecker Denschlag, Coherent Optical Transfer of Feshbach Molecules to a Lower Vibrational State, Phys. Rev. Lett. **98**, 043201 (2007).

[Wynar 2000] R. Wynar, R. S. Freeland, D. J. Han, C. Ryu, and D. J. Heinzen, Molecules in a Bose-Einstein Condensate, Science 287, 1016 (2000).

[Xu 2003] K. Xu, T. Mukaiyama, J. R. Abo-Shaeer, J. K. Chin, D. E. Miller, and W. Ketterle, Formation of Quantum-Degenerate Sodium Molecules , Phys. Rev. Lett. **91**, 210402 (2003).

[Yelin 2006] S. F. Yelin, K. Kirby, and R. Cote, Schemes for robust quantum computation with polar molecules, Phys. Rev. A **74**, 050301(R) (2006).

[Zaccanti 2009] M. Zaccanti, B. Deissler, C. DÉrrico, M. Fattori, M. Jona-Lasinio, S. Müller, G. Roati, M. Inguscio, and G. Modugno, Observation of an Efimov spectrum in an atomic system, Nature Phys. **5**, 586 (2009).

[Zare 1973] R. N. Zare, A. L. Schmeltekopf, W. J. Harrop, and D. L. Albritton, A direct approach for the reduction of diatomic spectra to molecular constants for the construction of RKR potentials, J. Mol. Spectrosc. **46**, 37 (1973).

[Zahzam 2006] N. Zahzam, T. Vogt, M. Mudrich, D. Comparat, P. Pillet, Atom-

Molecule Collisions in an Optically Trapped Gas, Phys. Rev. Lett. **96**, 023202 (2006).

[Zelevinsky 2008] T. Zelevinsky, S. Kotochigova and J. Ye, Precision Test of Mass-Ratio Variations with Lattice-Confined Ultracold Molecules, Phys. Rev. Lett. **100**, 043201 (2008).

[Zwierlein 2003] M. W. Zwierlein, C. A. Stan, C. H. Schunck, S. M. F. Raupach, S. Gupta, Z. Hadzibabic, and W. Ketterle, Observation of Bose-Einstein Condensation of Molecules, Phys. Rev. Lett. **91**, 250401 (2003).

# i want morebooks!

Buy your books fast and straightforward online - at one of world's fastest growing online book stores! Environmentally sound due to Print-on-Demand technologies.

## Buy your books online at
## www.get-morebooks.com

Kaufen Sie Ihre Bücher schnell und unkompliziert online – auf einer der am schnellsten wachsenden Buchhandelsplattformen weltweit! Dank Print-On-Demand umwelt- und ressourcenschonend produziert.

## Bücher schneller online kaufen
## www.morebooks.de

VDM Verlagsservicegesellschaft mbH
Heinrich-Böcking-Str. 6-8
D - 66121 Saarbrücken

Telefon: +49 681 3720 174
Telefax: +49 681 3720 1749

info@vdm-vsg.de
www.vdm-vsg.de

Printed by Books on Demand GmbH, Norderstedt / Germany